FLUID MECHANICS

FOR ENGINEERING TECHNOLOGY

PRENTICE-HALL INTERNATIONAL, INC., *London*
PRENTICE-HALL OF AUSTRALIA, PTY. LTD., *Sydney*
PRENTICE-HALL OF CANADA, LTD., *Toronto*
PRENTICE-HALL OF INDIA PRIVATE LTD., *New Delhi*
PRENTICE-HALL OF JAPAN, INC., *Tokyo*

FLUID MECHANICS

FOR ENGINEERING TECHNOLOGY

IRVING GRANET

New York Institute of Technology
Advisory Council for the Technologies,
Queensborough Community College

Prentice-Hall, Inc.

Englewood Cliffs, N.J.

Library of Congress Catalog Card Number: 79-114697
Printed in the United States of America

This book is dedicated to my devoted wife, Arlene, whose love and forebearance made its completion possible.

PREFACE

This book has been written for use in a course in fluid mechanics at the level of a student in either a technical institute or community college technology program. Based upon many years of association with students at this level, I have incorporated in the text the following features:

1. All physical principles are developed as required even though the student may have had a course in physics. A minimum of empirical data is used even though the text is oriented toward practical applications.

2. Since the students at this level may not have reached the point in their professional development where they can concurrently handle the Calculus while learning and developing new concepts, I have avoided the use of Calculus. In taking this approach, I feel that I am fulfilling the special need of the technology student for a more practical approach— one which completely avoids the Calculus but nevertheless gives the student the kind of background in fluid mechanics that will enable him to solve actual problems he will encounter in the field. My experience and the experience of my colleagues has been that the use of the "Δ" development yields a satisfactory approach. However, at the direction of the instructor, Calculus can be used for alternate parallel development to increase the depth of presentation where warranted and feasible.

3. The development of the generalized energy equation logically precedes the introduction of the Bernoulli equation. Therefore, I have taken care to develop the Energetics of Flow in some detail and from this general approach have developed the more restrictive Bernoulli equation. Again, this approach has been developed from my experience and is also based

upon "feedback" from former students and my colleagues. It is a logical and pedagogically sound approach.

4. The crowded curricula of these students usually permits time for only one course in fluid mechanics. I have, therefore, carefully selected topics that, with the appropriate emphasis by the instructor, makes this book appropriate for students majoring in Aeronautical, Civil, Mechanical, Aerospace, and Industrial Technology. In addition, I have incorporated a chapter on Fluidics which will serve as an introduction to this increasingly important branch of technology.

I am indebted to Professor Byron G. Schieber of Queensborough Community College for our long, professional friendship and association and for his insight and involvement with the problems of the technology student. President Kurt R. Schmeller of Queensborough Community College, in his vigorous approach to the formulation of programs that are up-to-date and responsive to the needs of both student and industry, has set the tone that I have tried to follow in this text. Additionally, I am grateful to Professor Allan Juster of the New York Institute of Technology who gave me the latitude that was necessary to develop and use much of the material in this text in my courses at the New York Institute of Technology.

Dr. William J. Guman and Dr. William McIlroy read and critically reviewed portions of the manuscript and Miss Ruby Ladislav and Mrs. Joan Fairfull typed both the rough and final drafts. My wife Arlene guided me in many ways on the pedagogical aspects of this text based upon her association with the Education Department of Hofstra University.

To all these good friends, colleagues, and my devoted wife, I wish to convey my heartfelt thanks. Finally, I am grateful to my children, Ellen, Kenneth, and David, who in the wisdom of their young years, gave me the necessary loving environment for this undertaking.

IRVING GRANET

North Bellmore, New York

CONTENTS

FLUID MECHANICS

FOR ENGINEERING TECHNOLOGY

Chapter 1

FUNDAMENTAL CONCEPTS

1.1 Introduction

Fluid mechanics is the study of the behavior of fluids whether they are at rest
or in motion; the study of fluids at rest is known as fluid statics and the study
of fluids in motion is termed fluid dynamics. In this book we shall use the
term *fluid* to refer to both gases and liquids. To distinguish between a liquid
and a gas, we shall note that while both will occupy the container in which
they are placed, a liquid presents a free surface if it does not completely fill
the container but a gas will always fill the volume of the container in which it
is placed. For gases it is important to take into account the change in volume
that occurs when either the pressure or temperature is changed, while in most
cases it is possible to neglect the change in volume of a liquid when there is a
change in pressure.

It is apparent that almost every part of our lives and the technology of
modern life involves some dependence on and a knowledge of the science of
fluid mechanics. Whether we consider the flow of blood in the minute blood

vessels of the human body or the motion of an aircraft or missile at speeds exceeding the velocity of sound, we shall need to utilize some branch of fluid mechanics to describe the motion. The literature of this subject is so vast that it is impossible to adequately briefly describe it in its entirety. At one time this subject was treated from a purely mathematical approach by one group of investigators and from an entirely empirical experimental approach by another group of investigators. In this text we shall use the modern technique of co-ordinating both approaches by supplementing theory with experiment.

1.2 Systems and Properties

As a general concept applicable to all situations we can simply define a system as a grouping of matter taken in any convenient or arbitrary manner. In the study of the behavior of the motion of solids, the student was introduced to the concept that a free-body diagram is a schematic portrayal of a body (or portion of a body) showing all of the external forces acting on it. Extensive use is made in Chapter 3 of both the system considerations and the free-body diagram, and we shall see that both of these concepts are indispensable tools in our study of fluid mechanics. For the present we shall simply note that systems can have energy stored or transferred to or from them, and although we are at liberty to choose a system in an arbitrary manner, all forces and energies must be accounted for when deriving the equations governing the system and its motion.

Let us consider a given system and then ask ourselves the question as to how we can distinguish changes that occur to it. To answer this question it is necessary for us to identify those external characteristics that will enable us to both distinguish and evaluate system changes. Conversely, if the external characteristics of a system do not change, then we should be able to infer that the system has not undergone any change. Some of the external characteristics that can be used to describe a system are temperature, pressure, volume, velocity, and position. These observable external characteristics of a system are called properties and we note that when all of the properties of a system are reproduced at different times and we are unable to distinguish any difference in the system, the system is in the same state at both times.

1.3 Temperature

The temperature of a system is a measure of the random motion of the molecules of the system. If there are different temperatures within the body (or bodies composing the system), the question arises as to how the temperature at a given location is measured and how this measurement is interpreted. Let us examine this question since similar questions will also have to be considered when other properties of a system are studied. In air at room pressure and temperature there are approximately 2.7×10^{19} molecules per cubic centimeter. If we divide the cube whose dimensions are one centimeter (1cm) on a side into smaller cubes each of whose sides is one thousandth of a centimeter, there will be about 2.7×10^{10} molecules in each of the smaller cubes, which is still an extraordinarily large number. Although we speak of temperature at a point, we really mean the average temperature of the molecules in the neighborhood of the point.

When measuring the temperature of a body with a thermometer it should be noted that the thermometer measures only the temperature of its sensing head. For the thermometer to be able to measure the temperature of a system it is necessary for both the thermometer and the system to be in thermal equilibrium. This concept follows from the Zeroth Law of Thermodynamics and the interested student is referred to any standard text on thermodynamics for a further discussion of this point. The measurement of temperature usually employs the measurement of some secondary characteristic (such as the height of a liquid in a bulb thermometer) when the system changes from one state to another. Other examples of physical effects that have been used to indicate temperature are the linear expansion of liquids and solids, the pressure change of a confined gas, the volume change of bodies, and the generation of voltages when dissimilar metals are placed in contact with each other. While it is generally assumed that the student is familiar with the common temperature scales, a brief summary will be given below.

The common scales of temperature are called, respectively, the Fahrenheit and Centigrade (Celsius) temperatures and are defined by using the ice point and boiling point of water at atmospheric pressure. In the Centigrade temperature scale, the interval between the ice point and the boiling point is divided into 100 equal parts. In addition, as shown in Figure 1.1, the Centigrade ice

°F	°C	°K	°R	
212	100	373	672	Atmospheric boiling point
32	0	273	492	Ice-point
−460	−273	0	0	Absolute zero

Figure 1.1

point is zero and the Fahrenheit ice point is 32. The conversion from one scale to the other is directly derived from Figure 1.1 and results in the following relations:

$$°C = \tfrac{5}{9}(°F - 32) \tag{1.1}$$

$$°F = \tfrac{9}{5}(°C) + 32 \tag{1.2}$$

The ability to extrapolate to temperatures below the ice point and above the boiling point of water and to interpolate in these regions is provided by the International Scale of Temperature. This agreed upon standard utilizes the boiling and melting points of different elements and establishes suitable interpolation formulas in the various temperature ranges between these elements. The data for these elements are given in Table 1.1.

Table 1.1

Element	Melting or Boiling Point at 1 Atm	Temperature	
		°C	°F
Oxygen	Boiling	−182.97	−297.35
Sulfur	Boiling	444.60	832.28
Antimony	Melting	630.50	1166.90
Silver	Melting	960.8	1761.4
Gold	Melting	1063.0	1945.4
Water	Melting	0	32
	Boiling	100	212

ILLUSTRATIVE PROBLEM 1.1. Determine the temperature at which the same value is indicated on both Fahrenheit and Centigrade thermometers.

Solution. Using equation (1.1) and letting $°C = °F$,

$$°F = \tfrac{5}{9}(°F - 32)$$

$$\tfrac{4}{9}°F = -\tfrac{160}{9}$$

$$°F = -40$$

Therefore, both Centigrade and Fahrenheit temperature scales indicate the same temperature at $-40°$.

By using the results of illustrative problem 1.1, it is possible to derive an alternative set of equations to convert from the Fahrenheit to the Centigrade temperature scale. When this is done, we obtain,

$$°F = \tfrac{9}{5}(40 + °C) - 40 \tag{1.3}$$

$$°C = \tfrac{5}{9}(40 + °F) - 40 \tag{1.4}$$

The symmetry of equations (1.3) and (1.4) makes them relatively easy to remember and use.

1.4 Absolute Temperature

Let us consider the case of a gas that is confined in a cylinder with a constant cross-sectional area by a piston that is free to move. If heat is now removed from the system, the piston will move down, but due to its weight it will maintain a constant pressure on the gas. This procedure can be carried out for several gases, and if volume is plotted as a function of temperature, we shall obtain a family of straight lines that intersect at zero volume (Figure 1.2a).

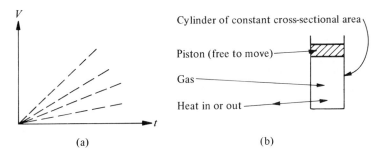

Figure 1.2. *Gas thermometer.*

This unique temperature is known as the absolute zero temperature, and the accepted values on the Fahrenheit and Centigrade temperature scales are $-459.69°$ and $-273.16°$, respectively, with the values $-460°$ and $-273°$ used for most engineering calculations. It is also possible to define an absolute temperature scale that is independent of the properties of any substance, and the interested student is referred to a text on thermodynamics for this development.

From the above, we have

$$\text{Degrees Rankine} = °R = °F + 460 \qquad (1.5)$$
$$\text{Degree Kelvin} = °K = °C + 273 \qquad (1.6)$$

The relation between degrees Rankine, degrees Fahrenheit, degrees Kelvin, and degrees Centigrade is shown graphically by Figure 1.1.

1.5 Pressure

When a gas is confined in a container, molecules of the gas strike the sides of the container, and these collisions with the walls of the container cause the molecules of the gas to exert a force on the walls. When the component of the force that is perpendicular (normal) to the wall is divided by the area of the wall, the resulting normal force per unit area is called pressure. To define the pressure at a point it is necessary to consider the area in question to be shrinking steadily. Pressure at a point is defined to be the normal force per unit area in the limit as the area goes to zero. Mathematically,

$$p = \left(\frac{\Delta F}{\Delta A}\right)_{\lim (\Delta A \to 0)} \tag{1.7}$$

In common engineering units, pressure is expressed as pounds force per square inch or pounds force per square foot, which are usually abbreviated as psi or psf. While these units are most frequently found in the literature, certain others are also used because of the manner in which the pressure measurements are made.

Most mechanical pressure gages, such as the Bourdon gage, measure the pressure above local atmospheric pressure. This pressure is called gage pressure and is measured in units of pounds force per square foot or pounds force per square inch, i.e., psfg or psig. The relation of gage pressure to absolute pressure is shown on Figure 1.3 and is:

Absolute pressure = Atmospheric pressure + Gage pressure   (1.8)

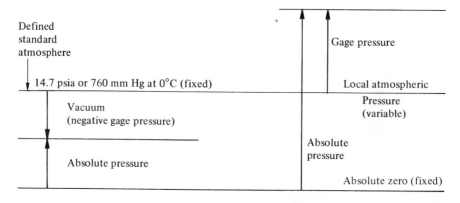

Figure 1.3

Absolute pressure is indicated as psia and psfa. It is usual to take the pressure of the atmosphere as being equal to 14.7 psia. However, in Europe 1 "ata" has been defined and is used as 1 kg/cm² for convenience, but this unit is not 14.7 psia; it is more nearly 14.2 psia. Care should be exercised when using the unit of pressure that is expressed as ata.

Referring again to Figure 1.3, it will be noted that pressure below atmospheric pressure is called vacuum. By reference to this figure we obtain the relation between vacuum and absolute pressure as

Vacuum = Atmospheric pressure − Absolute pressure   (1.9)

Density is defined as the mass per unit volume of a substance and *specific weight* is the weight per unit volume. In engineering units, density (ρ) is used as slugs per cubic foot and specific weight (γ) is in pounds per cubic foot. They are related by $\gamma = \rho g$, where g is the local acceleration of gravity.

We define *specific volume* as the reciprocal of specific weight or the volume per unit weight. The *weight* of a body is the force exerted on it by an external body such as the earth. Last, *specific gravity* is defined as the ratio of the weight of a body to the weight of an equal volume of water at a specific standard temperature.

ILLUSTRATIVE PROBLEM 1.2. Assume that a fluid whose specific weight is constant and equal to γ lb/ft³ is placed in a uniform tube to a height h above the base of the column. Determine the pressure at the base of the column.

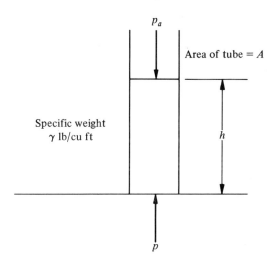

Solution. As is shown in the accompanying figure, the force on the base of the column is due to both the atmosphere and the force exerted by the weight of fluid. The volume of fluid in the tube is

$$V = Ah$$

and the weight of fluid is γ lb/ft³ × V ft³. Therefore, the weight is

$$W = Ah\gamma$$

In addition there is a force due to the atmosphere: $(p_a)(A)$, where p_a is the pressure of the atmosphere in pounds force per unit area. The total force on the base is

$$p_a A + Ah\gamma$$

From equation (1.7) the pressure on the base is therefore

$$p = \frac{p_a A + Ah\gamma}{A} = p_a + \gamma h$$

This is an absolute pressure. The pressure due only to the fluid is just $p = \gamma h$, where γ is the specific weight of the fluid in the tube, and

$$p = \rho g h \qquad \text{since } \gamma = \rho g \tag{1.10}$$

The most common fluid used to measure pressure differences, atmospheric pressure, and vacuum pressure is mercury. At approximately room temperature the specific weight of mercury is very nearly 13.6 g/cm³. It will usually be assumed that the specific weight of mercury is 13.6 g/cm³ unless otherwise stated. The effect of temperature on the specific weight of mercury is given in Appendix B.

ILLUSTRATIVE PROBLEM 1.3. A tube of glass open at the top has a 1-in. level of mercury in it. Take the specific weight of mercury (Hg) to be 13.6 g/cm³ and determine the pressure on the base of the column.

Solution. To solve this problem let us first convert the unit of weight of grams per cubic centimeter to pounds per cubic foot. In one pound there are approximately 454 grams. Also, there are 2.54 centimeters in 1 inch. Thus 1 gram is $\frac{1}{454}$ pound and one foot has 12×2.54 centimeters. To convert one gram per cubic centimeter to pounds per cubic foot we can apply the following dimensional reasoning:

$$\frac{g}{cm^3} \times \frac{lb}{g} \times \frac{cm^3}{ft^3} = \frac{lb}{ft^3} = \gamma$$

The use of dimensional reasoning (as well as the definition of the word *dimension*) is treated in more detail in Chapter 5.
Continuing,

$$\frac{1\,g}{cm^3} \times \frac{1}{454}\frac{lb}{g} \times (2.54 \times 12)^3 \frac{cm^3}{ft^3} = \frac{62.4\,lb}{ft^3} = \gamma$$

The conversion from grams per cubic centimeter pounds per cubic foot is thus 62.4. The student is strongly urged to check all computations for dimensional consistency when doing problems. Chapter 5 on dimensional homogeneity and dimensional analysis further emphasizes this point.
Applying equation (1.10),

$$p = \left(\frac{1}{12}\,ft\right)(13.6 \times 62.4)\frac{lb}{ft^3} = \frac{lb}{ft^2}$$

In pounds force per square inch,

$$p = \frac{1}{12} \times 13.6 \times 62.4 \times \frac{1}{144} \frac{ft^2}{in.^2} = 0.491 \text{ psi}$$

This is a gage pressure. If the local atmospheric pressure is 14.7 psia,

$$p = 0.491 + 14.7 = 15.19 \text{ psia}$$

The value of 0.491 psi/in. Hg is a useful conversion factor.

In vacuum work it is common to express the absolute pressure in a vacuum chamber in terms of millimeters of mercury. Thus a vacuum may be expressed as 10^{-5} mm Hg. If this is expressed in pounds force per square inch, it would be equivalent to 0.000000193 psi, with the assumption that the density of mercury (for vacuum work) is 13.6 g/cm^3. Recently, the term *torr* has entered the technical literature. A torr is defined as 1 mm Hg. Thus 10^{-5} torr is the same as 10^{-5} mm Hg. Another unit of pressure used in vacuum work is the micron. A *micron* is defined as one thousandth of 1 mm Hg so that 10^{-3} mm Hg is equal to 1 micron.

ILLUSTRATIVE PROBLEM 1.4. A mercury manometer (vacuum gage) reads 26.5 in. of vacuum when the local barometer reads 30.0 in. Hg at standard temperature. Determine the absolute pressure in psia.

Solution

$p = (30.0 - 26.5)$ in. Hg vacuum, and from the proceding problem

$p = (30 - 26.5)0.491 = 1.72$ psia

If the solution is desired in psfa, it is necessary to convert from square inches to square feet. Since there are 12 in. in 1 ft, there will be 144 in^2. in 1 ft^2. Thus the conversion to psfa requires that psia be multiplied by 144. This conversion is often necessary to keep the dimensions of equations consistent.

One word of caution must be expressed at this point. The word *pound* has been used interchangeably for both force and mass. In using a consistent set of units it is possible to use the word *pound* for both force and mass, and this is commonly done in texts on thermodynamics. In the field of fluid mechanics it is more customary to use the American system, which employs the pound as the unit of force and the slug as the unit of mass. When this later system is used, the conversion factor relating mass and force is found from Newton's second law: Force equals the product of mass and acceleration ($F = Ma$). Thus a pound force acting on a mass of 1 slug will accelerate it at a rate of 1 ft/sec/

sec. Mathematically, a pound force is equivalent to a mass of 1 slug multiplied by the local acceleration of gravity. On the earth the acceleration of gravity can be taken as 32.2 ft/sec² for most engineering calculations.

1.6 Viscosity

Because of the attractive forces between their molecules, fluids have the ability to resist forces tending to change the shape of a body of fluid. The stresses within an elemental volume of fluid are analogous to those within an elastic body. In an elastic body the stresses are proportional to the change of shape of the body; i.e., stress is proportional to strain. In the case of fluids the attractive forces between molecules are weaker, and so stress is assumed to be proportional to the time rate of change of shape with respect to distance; the constant of proportionality is called viscosity. This assumption was introduced by Newton, who set the force acting tangentially on a unit surface between adjacent layers of fluid (i.e., the stress) equal to a constant (i.e., viscosity) times the velocity gradient in a direction perpendicular to the layers.

Consider the situation shown in Figure 1.4. The upper plate is moving to

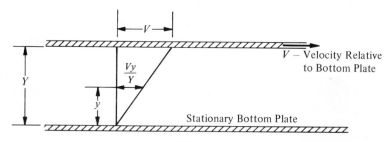

Figure 1.4. Viscosity.

the right with a velocity V relative to the lower plate. If the viscosity is constant, the variation of velocity between the plates will be linear. The particle in contact with the bottom plate will not be moving, while the particle touching the upper plate will be moving with the velocity of the upper plate, V. As shown, the velocity anywhere between the plates will be simply given by Vy/Y. The force required to move the upper plate at constant velocity relative to the bottom plate is proportional to the area of the plates, inversely proportional to the distance separating the plates, and directly proportional to the relative velocity of the plates. Thus,

$$F \propto \frac{V}{Y} A$$

By definition, stress is force divided by area. In this case the stress is a shear stress in that the fluid acts as if each layer were composed of thin sheets tending to move (shear) with respect to each other. The shear stress, τ, is therefore

$$\tau \propto \frac{V}{Y} \tag{1.11}$$

Expressing the proportionality of equation (1.11) as an equality,

$$\frac{F}{A} = \tau = \mu \frac{V}{Y} \tag{1.12}$$

The proportionality constant μ is known as the *coefficient of vicosity*, *dynamic viscosity*, the *absolute viscosity*, or simply the *viscosity of a fluid*. If the viscosity varies with y due to temperature or other local conditions, μ can be written as,

$$\mu = \tau \left(\frac{\Delta y}{\Delta V}\right)_{\lim(\Delta V \to 0)} \tag{1.13}$$

where μ may vary from location to location.

Since shear stress has the dimensions of force divided by area, and velocity has the dimensions of length per unit time, viscosity must have the dimension of force per unit area multiplied by time. The units of viscosity are therefore pound seconds per square foot, gram seconds per square centimeter, poundal seconds per square foot, dyne seconds per square centimeter, or dyne seconds per square foot. Of these units, dyne seconds per square centimeter is given the name *poise* and it is found that the viscosity of water at 68.4°F is 1 cP or one hundredth of 1 P. The viscosity of most liquids and gases is essentially independent of pressure except for very high pressures, but as temperature increases, the viscosity of all liquids decreases, while the viscosity of all gases increases, as can be seen from Figure 1.5. Table 1.2 gives some convenient conversion factors for viscosity.

Table 1.2

1 P $=$ 0.0672 poundal sec/ft^2
1 P $=$ 0.00209 lb sec/ft^2
1 poundal sec/ft^2 $=$ 14.88 P
1 lb sec/ft^2 $=$ 478.8 P
1 cm^2/sec $=$ 0.001076 ft^2/sec
1 ft^2/sec $=$ 929 cm^2/sec

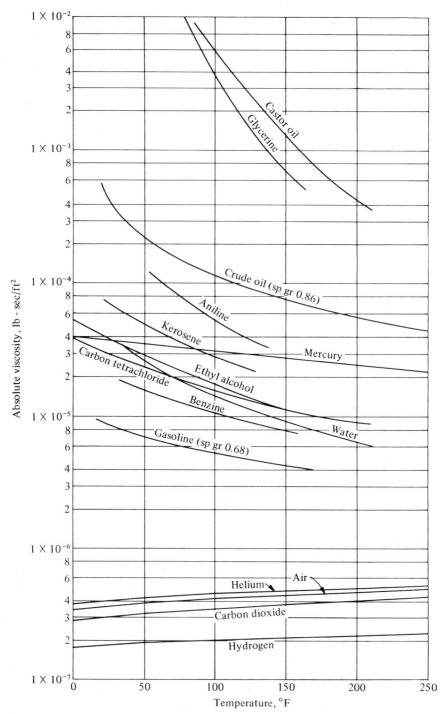

Figure 1.5. Absolute viscosities of certain gases and liquids. Reproduced with permission from Fluid Mechanics by V. L. Streeter, McGraw-Hill Book Company, Inc., New York, 1962, p. 534.

ILLUSTRATIVE PROBLEM 1.5. Derive the conversion factor that 1 P = 0.00209 lb sec/ft².

Solution. To solve a problem of this type it is first necessary to define all terms. A poise is a dyne second per square centimeter and the dyne is given in the centimeter-gram-second system as a gram centimeter per second squared. Therefore

$$1\ P = \frac{1\ \text{dyne sec}}{\text{cm}^2} = \left[\left(\frac{1\ \text{dyne sec}}{\text{cm}^2}\right)\right]\frac{1}{\dfrac{1\ \text{ft}^2}{(12 \times 2.54)^2\,\text{cm}^2}}$$

$$= \frac{1\ \text{dyne sec}}{\text{ft}^2}$$

$$1\ \text{dyne} = \frac{1\ \text{g cm}}{\text{sec}^2} = \left(\frac{1\ \text{g cm}}{\text{sec}^2}\right) \times \frac{1}{980\dfrac{\text{cm}}{\text{sec}^2}}\,\frac{1}{454\dfrac{\text{g}}{\text{lb}}} = \text{lb}$$

Therefore

$$1\ P = \left[\frac{1}{\dfrac{1}{(12 \times 2.54)^2}} \times \frac{1}{980 \times 454}\right] = \frac{\text{lb sec}}{\text{ft}^2}$$

$$= 0.00209\ \frac{\text{lb sec}}{\text{ft}^2}$$

It is important to keep in mind at all times that consistent sets of units must be used in all problems. The extra time required to check this point is relatively small when we find that mistakes and inconsistencies in units and conversions are the greatest single source of errors in problems in this subject.

It is sometimes convenient to combine viscosity and density; viscosity divided by density is known as *kinematic viscosity*. In the metric system the name stoke (or one hundredth of its value, the centistoke) is used to designate kinematic viscosity; a centistoke is 1 cP divided by a density of 1 g/cm³. Figure 1.6 shows the kinematic viscosity of certain gases and liquids, where the gases are at atmospheric pressure.

ILLUSTRATIVE PROBLEM 1.6. The viscosity of mercury is 33×10^{-6} lb sec/ft² at 68°F. Determine the force necessary to maintain a relative velocity of 5 ft/sec between two plates that are separated by 4 in. and whose area is 1 ft². Consider only viscous effects.

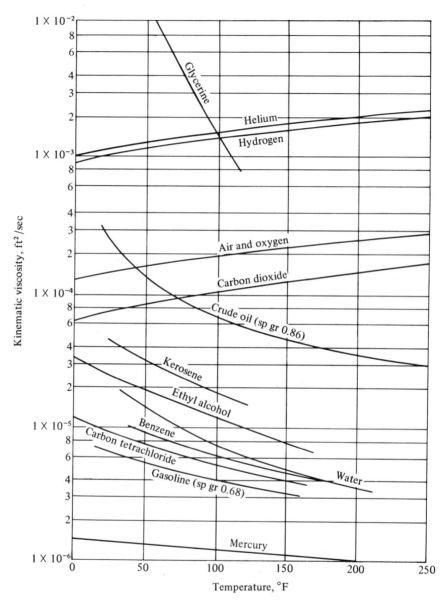

Figure 1.6. *Kinematic viscosities of certain gases and liquids. Reproduced with permission from* Fluid Mechanics *by V. L. Streeter, McGraw-Hill Book Company, Inc., New York, 1962., p. 535.*

Solution. From equation (1.12),

$$\frac{F}{A} = \mu \frac{V}{Y}$$

$$\therefore F = \mu A \frac{V}{Y} = \frac{33 \times 10^{-6} \times 1 \times 5}{\frac{1}{3}} = 495 \times 10^{-6} \, lb$$

1.7 Surface Tension

When a liquid has a free surface there is a discontinuity that takes place at the surface since there is a liquid on one side of the surface and a different liquid and gas on the other side of the surface. For example, consider a drop of liquid in air. Within the drop, each molecule of liquid experiences forces due to the other liquid molecules. At the surface of the drop the liquid molecules experience forces due to both the air and the liquid molecules. If the drop is to maintain its shape, the attraction of the liquid molecules must exceed the attraction between the liquid molecules and air molecules. Effectively, this is the same as if the liquid were enclosed by a stretchable membrane under tension. The force of this *surface tension* is always tangent to the interface. A liquid is said to wet a surface in contact with it if the attraction of the molecules for the surface exceeds the attraction of the molecules for each other; on the other hand, the liquid is classified as a nonwetting liquid if the attraction of other liquid molecules for each other is greater than their attraction to the surface.

Quantitatively, surface tension can be defined as the force of molecular attraction per unit length of free surface. It should be noted that surface tension is a function of both the liquid and the surface in contact with the liquid. Denoting surface tension by σ, we have, in general,

$$\sigma = \frac{F}{L} \tag{1.14}$$

Since the forces of attraction between molecules decreases with increasing temperature, it is found that surface tension also decreases with increasing temperature. Table 1.3 gives the surface tension of some fluids in contact with air at $68\,^\circ F$.

With the foregoing in mind, let us refer to Figure 1.7, which shows the contact angle relations for the solid-liquid-vapor interfaces for both wetting and nonwetting liquids. The subscripts *vs*, *vl*, and *ls* denote, respectively, the vapor-solid interface, the vapor-liquid interface, and the liquid-solid interface. For the liquid to be in equilibrium with the solid surface the surface tension

Table 1.3† Surface Tension of Common Liquids in Contact with Air at 68°F‡

	σ (lb/ft)
Alcohol, ethyl	0.00153
Benzene	0.00198
Carbon tetrachloride	0.00183
Kerosene	0.0016–0.0022
Water	0.00498
Mercury	
In air	0.0352
In water	0.0269
In vacuum	0.0333
Oil	
Lubricating	0.0024–0.0026
Crude	0.0016–0.0026

† Reproduced with permission from *Fluid Mechanics* by V. L. Streeter, McGraw-Hill Book Company, Inc., New York, 1958, p. 14.
‡ See Tables C.7 and C.8 in Appendix C for other data on surface tension.

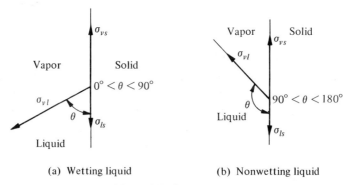

(a) Wetting liquid   (b) Nonwetting liquid

Figure 1.7. Contact angles.

forces at the solid-liquid-vapor interfaces must be in balance parallel to the solid surface. Therefore

$$\sigma_{vs} = \sigma_{ls} + \sigma_{vl} \cos \theta \tag{1.15}$$

where θ is the contact angle between the fluid and the surface. When $(\sigma_{vs} - \sigma_{ls})/\sigma_{vl}$ is between zero and unity, the contact angle lies between 0 and 90 deg and the liquid is called a wetting liquid. If $(\sigma_{vs} - \sigma_{ls})/\sigma_{vl}$ lies between 0 and -1, the contact angle will be between 90 and 180 deg and the liquid is termed nonwetting. The magnitude of the contact angle θ is dependent on the three surface tension forces.

Let us calculate the rise h shown in Figure 1.8a for the case of a wetting

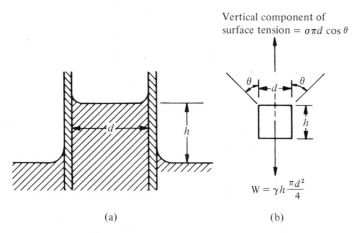

Vertical component of
surface tension = $\sigma \pi d \cos \theta$

$$W = \gamma h \frac{\pi d^2}{4}$$

(a)   (b)

Figure 1.8. *Surface tension—wetting liquid.*

fluid in a circular tube. The basic consideration here is that the weight of the column of height h must be supported by an equal force in the vertical direction where the source of this supporting force must be the vertical component of the surface tension. Therefore the force down (equal to the weight of the column of height h) is $\gamma h(\pi d^2/4)$, where γ is the specific weight of the fluid and d is the inside diameter of the tube. The total force due to the surface tension is the surface tension multiplied by the perimeter of the tube: $\sigma \pi d$. However, since we only desire the vertical component of this force, it is necessary to multiply the total force by $\cos \theta$. Equating the two vertical forces,

$$\gamma h \frac{\pi d^2}{4} = \sigma \pi d \cos \theta$$

or

$$h = \frac{4\sigma \cos \theta}{\gamma d} = \frac{4\sigma \cos \theta}{\rho g d} \qquad (1.16)$$

Water in contact with glass and air has a contact angle whose value is essentially zero, while mercury in contact with glass and air has a contact angle of 129 deg.

If we consider two infinite parallel plates instead of a circular tube, the equation of equilibrium becomes (for a unit depth)

$$\gamma h d = 2\sigma \cos \theta$$

or

$$h = \frac{2\sigma \cos \theta}{\gamma d} = \frac{2\sigma \cos \theta}{\rho g d} \tag{1.17}$$

Notice that the rise in the tube is twice the rise between the parallel plates.

ILLUSTRATIVE PROBLEM 1.7. A glass tube 0.005 ft i.d. is placed in a fluid whose contact angle is 10 deg. If the surface tension for this fluid in contact with glass is 5×10^{-3} lb/ft, determine the rise in the tube above the normal liquid level. Assume that the specific weight of the fluid is 50 lb/ft³.

Solution. From equation (1.16),

$$h = \frac{4\sigma \cos \theta}{\gamma d} = \frac{4 \times 5 \times 10^{-3} \cos 10°}{50 \times 0.005} = 0.0798 \text{ ft} = 0.96\text{-in. rise}$$

1.8 Compressibility

When a liquid is confined it will offer resistance to any external agency that tends to change its shape. This resistance to change in shape is called the compressibility of the fluid, and the change in volume is measured as a function of the original volume and the applied pressure. Mathematically, the *modulus of elasticity* or *bulk modulus* of the fluid at a point is defined by

$$\beta = -V \left(\frac{\Delta p}{\Delta V} \right)_{\lim (\Delta V \to 0)} \tag{1.18}$$

where the negative sign accounts for the fact that as the pressure increases the volume decreases. It is usually assumed that the bulk modulus is insensitive to pressure but does vary with temperature. For water the bulk modulus is approximately 300,000 lb/in². Since steel has a modulus of elasticity of 30 million psi it is seen that water is relatively compressible. The physical properties of water are given in Table 1.4 as a function of temperature. Figure C.1 gives the variation of the bulk modulus of water with pressure and temperature.

ILLUSTRATIVE PROBLEM 1.8. Determine the percentage of change in the volume of water if the pressure is increased by 30,000 psi.

Solution. From equation (1.18),

$$\beta = -V\left(\frac{\Delta p}{\Delta V}\right)_{\lim (\Delta V \to 0)}$$

or

$$\frac{\Delta V}{V} = -\left(\frac{\Delta p}{\beta}\right) = \frac{-30,000}{300,000} = -0.1 \quad \text{or} \quad -10\%$$

Table 1.4† Physical Properties of Water

Temperature, ($°F$)	Specific Weight, γ (lb/ft^3)	Density, ρ ($slugs/ft^3$)	Viscosity, μ ($lb\ sec/ft^2$) $10^5\mu =$	Kinematic Viscosity, ν (ft^2/sec) $10^5\nu =$	Surface Tension, σ (lb/ft) $100\sigma =$	Vapor Pressure Head, p_v/γ (ft)	Bulk Modulus of Elasticity, β ($lb/in.^2$) $10^{-3}\beta =$
32	62.42	1.940	3.746	1.931	0.518	0.20	293
40	62.43	1.940	3.229	1.664	0.514	0.28	294
50	62.41	1.940	2.735	1.410	0.509	0.41	305
60	62.37	1.938	2.359	1.217	0.504	0.59	311
70	62.30	1.936	2.050	1.059	0.500	0.84	320
80	62.22	1.934	1.799	0.930	0.492	1.17	322
90	62.11	1.931	1.595	0.826	0.486	1.61	323
100	62.00	1.927	1.424	0.739	0.480	2.19	327
110	61.86	1.923	1.284	0.667	0.473	2.95	331
120	61.71	1.918	1.168	0.609	0.465	3.91	333
130	61.55	1.913	1.069	0.558	0.460	5.13	334
140	61.38	1.908	0.981	0.514	0.454	6.67	330
150	61.20	1.902	0.905	0.476	0.447	8.58	328
160	61.00	1.896	0.838	0.442	0.441	10.95	326
170	60.80	1.890	0.780	0.413	0.433	13.83	322
180	60.58	1.883	0.726	0.385	0.426	17.33	318
190	60.36	1.876	0.678	0.362	0.419	21.55	313
200	60.12	1.868	0.637	0.341	0.412	26.59	308
212	59.83	1.860	0.593	0.319	0.404	33.90	300

† Reproduced with permission from *Fluid Mechanics* by V. L. Streeter, McGraw-Hill Book Company, Inc., New York, 1962, p. 533.

1.9 Closure

Fluid mechanics is a vast field that has engaged the combined efforts of some of the greatest, scientists, engineers, mathematicians, and physicists. However, in our introductory study of this field we shall naturally be limited to certain topics. In Chapter 1 certain basic concepts have been developed that will be useful as this study progresses, and the student should master these before proceeding with the following chapters. Several items have been noted that are important and that are briefly repeated here:

1. Be sure of the units used in an equation.
2. Draw a free-body diagram where possible.
3. It is important to keep in mind the definitions of all terms used.

Reading and understanding text material is a necessity; mastery of the material is an accomplished fact when the student can apply it to the solution of problems. A large number of illustrative problems are included in all of the chapters to illustrate text material. The student is urged to solve as many of these as possible without reference to the text to evaluate his level of understanding. Problems at the end of the chapters also serve to establish a better level of understanding through the application of the material given within each chapter.

REFERENCES

1. *Fluid Mechanics for Engineers* by P. S. Barna, Butterworth & Co. (Publishers) Ltd., London, 1957.

2. *Elementary Applied Thermodynamics* by I. Granet, John Wiley & Sons, Inc., New York, 1965.

3. *Elementary Theoretical Fluid Mechanics* by K. Brenkert, Jr., John Wiley & Sons, Inc., New York, 1960.

4. *Mechanics of Fluids* by G. Murphy, 1st ed., International Textbook Company, Scranton, Pa., 1942.

5. *Engineering Applications of Fluid Mechanics* by J. C. Hunsaker and B. G. Rightmire, McGraw-Hill Book Company, Inc., New York, 1947.

6. *Fluid Mechanics* by R. C. Binder, 4th ed., Prentice-Hall, Inc., Englewood Cliffs, N. J., 1962.

7. *Thermodynamics of Fluid Flow* by N. A. Hall, Prentice-Hall, Inc., Englewood Cliffs, N. J., 1951.

8. *Hydraulics* by R. L. Daugherty, 4th ed., McGraw-Hill Book Company, Inc., New York, 1937.

9. *Fluid Mechanics* by V. L. Streeter, 3rd ed., McGraw-Hill Book Company, Inc., New York, 1958, 1962.

10. *Fundamentals of Hydro- and Aero-Mechanics* by L. Prandtl and O. G. Tietjens, McGraw-Hill Book Company, Inc., New York, 1934.

11. *Elementary Fluid Mechanics* by J. K. Vennard, John Wiley & Sons, Inc., New York, 1961.
12. *A Textbook of Fluid Mechanics* by J. R. D. Frances, Edward Arnold & Co., London, 1958.

PROBLEMS

1.1 Convert 20°, 40°, and 60°C to equivalent degrees Fahrenheit.

1.2 Change 0°, 10°, and 50°F to equivalent degrees Centigrade.

1.3 Convert 500°R, 500°K, and 650°R to degrees Centigrade.

1.4 An arbitrary temperature scale is proposed in which 20° is assigned to the ice point and 75° is assigned to the boiling point. Derive an equation relating this scale to the Centigrade scale.

1.5 For the temperature scale proposed in problem 1.4, what temperature corresponds to absolute zero?

1.6 Derive a relation between degrees Rankine and degrees Kelvin, and based upon the results, show that $(°C + 273)1.8 = °F + 460$.

1.7 Derive equations (1.3) and (1.4).

1.8 A pressure gage indicates 25 psi when the barometer is at a pressure equivalent to 14.5 psia. Compute the absolute pressure in psi and feet of mercury when the specific weight of mercury is $13.0 \ g/cm^3$.

1.9 The same as problem 1.8: The barometer stands at 750 mm Hg and its specific weight is $13.6 \ g/cm^3$.

1.10 A vacuum gage reads 8 in. Hg when the atmospheric pressure is 29.0 in. Hg. If the specific weight of mercury is $13.6 \ g/cm^3$, compute the absolute pressure in psi.

1.11 A column of fluid is 25 in. high. If the specific weight of the fluid is $60.0 \ lb/ft^3$, what is the pressure in psi at the base of the column?

1.12 Determine the density and specific volume of the contents of a 10-ft³ tank if the contents weigh 250 lb.

1.13 A new thermometer scale on which the ice point of water at atmospheric pressure would correspond to a marking of 200 and the boiling point of water at atmospheric pressure would correspond to a marking of −400 is proposed. What would be the reading of this new thermometer if a Fahrenheit thermometer placed in the same environment read 80°F?

1.14 Derive the conversion from poise to poundal seconds per square foot. Check the result with Table 1.2 (1 P = 0.0672 poundal sec/ft²).

1.15 Two plates separated by 3 in. are placed in oil with a viscosity of 133 × 10⁻⁶ lb sec/ft². If the plates are parallel and have an area of 5 ft², determine the relative velocity (considering only viscous effects) if a force of 0.1 lb is applied to one of the plates parallel to the orientation of the plates.

1.16 A tube whose diameter is 0.001 ft is placed in a mercury pool. If the surface tension of mercury is 35 × 10⁻³ lb/ft, determine the depression of the mercury in the tube below the surface of the mercury pool. Assume that $\theta = 129$ deg and that the specific weight of mercury is 13.6 times that of water.

1.17 If instead of the tube in problem 1.16 two infinite parallel plates are used, determine the depression of the mercury between the plates below the surface of the pool.

1.18 Determine the bulk modulus of a fluid if it undergoes a 1 percent change in volume when subjected to a pressure of 10,000 psi.

NOMENCLATURE

A = area
d = diameter
F = force
g = local acceleration of gravity
h = height
L = length of free surface
p = pressure
psia = pounds per square inch absolute
psig = pounds per square inch gage
V = volume
V = velocity
Y = linear plate separation
y = distance from a plate
$°C$ = degrees Centigrade
$°F$ = degrees Fahrenheit
$°K$ = degrees Kelvin

$°R$ = degrees Rankine
β = bulk modulus
γ = specific weight
Δ = small change or increment
θ = angle
μ = viscosity
ρ = density
σ = surface tension
τ = shear stress
ν = kinematic viscosity

Subscripts
a = atmosphere
ls = liquid-solid interface
vl = vapor-liquid interface
vs = vapor-solid interface

Chapter 2

FLUID
STATICS

2.1 Introduction

The study of fluids at rest or having no velocity with respect to an observer in a gravitational field is known as fluid statics. Since the fluid will be at rest with respect to an observer, there will be no relative motion between adjacent fluid layers and viscosity will not enter into problems in fluid statics. These considerations enable us to treat fluids at rest mathematically, and the results of these mathematical studies will be very accurate for engineering purposes. For most engineering purposes if a fluid is at rest with respect to a system that is moving with a uniform translation relative to the earth, it can be treated as if it were at rest since it will have no acceleration with respect to the earth.

To solve problems in fluid statics it is only necessary to use the principle of equilibrium in bodies that is developed in mechanics. Thus for a body (or element of mass) to be in equilibrium it is necessary that the sum of the external forces and moments acting on the body be zero. This principle, combined with a knowledge of the fluid density, enables one to readily analyze problems in fluid statics.

2.2 Pressure Relationships

In Chapter 1 the pressure at the base of a uniform column of liquid was derived as a function of the specific weight of the liquid and the height of the liquid column. The derivation was somewhat simplified, and it will be both useful and more general to look at this problem from an alternative viewpoint. Consider a disk-shaped element of fluid located a distance x from the base (or reference plane). The height of the column will be measured as positive in the vertical direction, as shown in Figure 2.1. The height of the disk will be

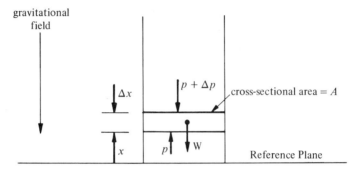

Figure 2.1. *Liquid-column analysis.*

Δx, and its cross-sectional area will be denoted as A. On the top of the disk there will be assumed to be a pressure $p + \Delta p$, where Δp is a small negative pressure increment. On the bottom of the disk it will be assumed that the pressure is slightly greater than this by an amount Δp. Thus on the bottom face of the disk the pressure is p. The weight of fluid is the volume of the disk $(\Delta x)A$ multiplied by the specific weight γ. Therefore, acting vertically down, we have the forces $(p + \Delta p)A + (A)(\Delta x \gamma)$, and acting vertically up, the force is pA. Equating forces in the vertical direction for the necessary condition of equilibrium yields,

$$pA = (p + \Delta p)A + A(\Delta x)(\gamma) \tag{2.1}$$

Simplifying equation (2.1) gives the result;

$$\frac{\Delta p}{\Delta x} = -\gamma \tag{2.2}$$

Equation (2.2) can be interpreted by referring to Figure 2.2, which is a plot of pressure against height. As can be seen, γ is simply the constant slope,

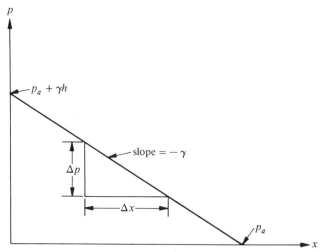

Figure 2.2. *p−x relation for a column of liquid.*

and from Figure 2.2 we can write directly

$$p = p_a + \gamma h - \gamma x = p_a + \gamma(h - x) \tag{2.3}$$

where p_a is the pressure at the free surface on top of the column. Equation (2.2) is the fundamental relation between pressure, specific weight, and column height. The negative sign indicates that the pressure decreases as one goes up the column. Note that the pressure in a column can be specified by stating feet of a fluid of a given density. In the nomenclature of hydraulics, this was known as a "head," and this nomenclature is still used today. Thus 3 ft of water represents the head of water or a gage pressure of $3 \times 62.4 = 187.2$ psfg or 1.3 psig.

If the liquid does not have a constant density (say due to temperature gradients or to pressure effects), it is necessary to evaluate equation (2.2) taking into account the fact that γ is a variable. This can be done by either using the methods of calculus or by graphically solving the equation. One way of evaluating this equation graphically is to evaluate the mean or average value of γ for a specified variation of γ as a function of height. Illustrative problem 2.1 demonstrates this procedure.

ILLUSTRATIVE PROBLEM 2.1. If γ varies linearly from γ_1 at the bottom of a column of fluid to γ_2 at the top of the column, determine the pressure at the bottom of the column.

Solution. A plot of γ against height is shown in Figure 2.3. By definition, the "average" is that value of γ that when multiplied by the base of the figure yields the area under the curve of the function. Therefore the area of the tri-

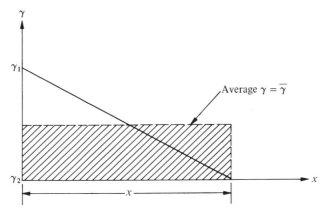

Figure 2.3. *Solution to illustrative problem 2.1.*

angle must equal the area of the shaded rectangle shown in the figure. Since the area of a triangle is half the base multiplied by the altitude,

$$\frac{x(\gamma_1 + \gamma_2)}{2} = \bar{\gamma}x$$

or

$$\bar{\gamma} = \frac{\gamma_1 + \gamma_2}{2}$$

Therefore the average γ, $\bar{\gamma}$, is simply the arithmetic mean of γ_1 and γ_2 *for this case only.* This corresponds to a uniform column of liquid of specific weight $(\gamma_1 + \gamma_2)/2$ in a tube.

There is one case of variable density that is of special use. Consider a condition where the local atmospheric pressure at sea level is known. If an airplane is at 1000 ft above sea level, what would the pressure be? Conversely, if the reading of a barometer is known at sea level and some unknown elevation what is the elevation? To answer this problem we shall make the assumption that the air temperature is constant and apply equation (2.2):

$$\frac{\Delta p}{\Delta x} = -\gamma \tag{2.2}$$

The relation that governs the pressure, temperature, and density for an *ideal gas* (air can be so considered for our purposes) is discussed in Chapter 7. It is

$$pv = RT \qquad \text{or} \qquad \frac{p}{\gamma} = RT \tag{2.4}$$

where p is the pressure in psfa, R is a constant for a gas (53.3 for air in these engineering units), and T is the absolute temperature in degrees Rankine. Inserting γ from equation (2.4) into equation (2.2),

$$\frac{\Delta p}{\Delta x} = \frac{-p}{RT} \qquad (2.5)$$

or

$$\frac{\Delta p}{p} = \frac{-\Delta x}{RT} \qquad (2.6)$$

To obtain a solution to equation (2.6), it is necessary to sum this equation between the required pressure limits. Graphically, this process of summation can be illustrated by referring to Figure 2.4. Plotting $1/p$ against p gives us the

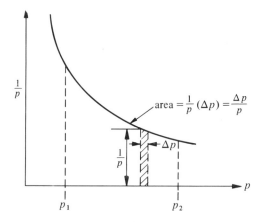

$$\text{area} = \frac{1}{p}(\Delta p) = \frac{\Delta p}{p}$$

Figure 2.4. Evaluation of $\Delta p/p$.

coordinates of this graph. By selecting a value of Δp as shown, the shaded area gives us $\Delta p/p$. Therefore, the sum of the $\Delta p/p$ values is the area under the curve between the limits of p_1 and p_2. By the methods of calculus, it can be shown that this area is $\ln(p_2/p_1)$, where $\ln x = \log_e x = 2.3026 \log_{10} x$.

Returning to equation (2.6) and noting that R and T are constants for this problem, we can easily evaluate the right-hand side graphically by plotting x against $1/RT$, as shown in Figure 2.5. For all values of Δx, $1/RT$ is a constant and the curve is simply a rectangle whose area is x/RT. Thus equation (2.4) becomes

$$\ln \frac{p_2}{p_1} = -\frac{x}{RT}$$

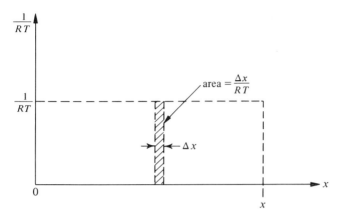

Figure 2.5. Evaluation of Δx/RT.

or

$$\ln \frac{p_1}{p_2} = \frac{x}{RT} \qquad (2.7)$$

The readings of a barometer are directly proportional to the local atmospheric pressure and therefore the ratio p_1/p_2 can be replaced by the barometer readings b_1 and b_2 as the ratio b_1/b_2. Equation (2.7) becomes

$$\ln \frac{b_1}{b_2} = \frac{x}{RT} \qquad (2.8)$$

where x is the difference in elevation and b_1 and b_2 are the barometer pressures at sea level and elevation, respectively.

ILLUSTRATIVE PROBLEM 2.2. A barometer reads 760 mm Hg at sea level and 750 mm Hg at some other elevation. If the air temperature can be taken to be constant and equal to 70 °F (530 °R), determine the unknown elevation.

Solution. To obtain accuracy in the solution of this problem, note that $\ln b_1/b_2 = \ln b_1 - \ln b_2$. Thus equation (2.8) becomes

$$\ln 760 - \ln 750 = \frac{x}{53.3 \times 530}$$

and

$$x = 53.3 \times 530(\ln 760 - \ln 750) = 53.3 \times 530 \times 0.01325 = 374 \text{ ft}$$

The student should note that for small elevation differences the assumption of constant temperature is quite good. For large elevation differences there is a large change in temperature and equation (2.7) is no longer applicable.

For convenience and for standardization of performance data, a *standard atmosphere* has been defined; it is given in Table 2.1. It will be noted from this table that the temperature is not constant and that the equations previously derived are valid only for small changes in elevation.

Table 2.1 The ICAO Standard Atmosphere[†]

Altitude (ft)	Temperature (°F)	Pressure (psia)	Specific Weight (lb/ft³)	Density (slug/ft³)	Viscosity × 10⁷ (lb sec/ft²)
0	59.00	14.696	0.07648	0.002377	3.737
5,000	41.17	12.243	0.06587	0.002048	3.637
10,000	23.36	10.108	0.05643	0.001756	3.534
15,000	5.55	8.297	0.04807	0.001496	3.430
20,000	−12.26	6.759	0.04070	0.001267	3.325
25,000	−30.05	5.461	0.03422	0.001066	3.217
30,000	−47.83	4.373	0.02858	0.000891	3.107
35,000	−65.61	3.468	0.02367	0.000738	2.995
40,000	−69.70	2.730	0.01882	0.000587	2.969
45,000	−69.70	2.149	0.01481	0.000462	2.969
50,000	−69.70	1.690	0.01165	0.000364	2.969
55,000	−69.70	1.331	0.00917	0.000287	2.969
60,000	−69.70	1.049	0.00722	0.000226	2.969
65,000	−69.70	0.826	0.00568	0.000178	2.969
70,000	−69.70	0.650	0.00447	0.000140	2.969
75,000	−69.70	0.512	0.00352	0.000110	2.969
80,000	−69.70	0.404	0.00277	0.000087	2.969
85,000	−65.37	0.318	0.00216	0.000068	2.997
90,000	−57.20	0.252	0.00168	0.000053	3.048
95,000	−49.05	0.200	0.00131	0.000041	3.099
100,000	−40.89	0.160	0.00102	0.000032	3.150

†Reproduced with permission from *Elementary Fluid Mechanics* by J. K. Vennard, McGraw-Hill Book Company, Inc., New York, 1961, p. 548.

If *h* is the height above sea level in feet and the temperature is assumed to decrease linearly, it can be approximated by

$$T = (519 - 0.00357h)\,°\text{R}$$

At 35,000 ft, the temperature becomes −67°F and is assumed to be constant above this level.

Let us now ask the question whether it is possible to have a pressure at a given horizontal level in a static fluid differ from point to point. To answer this question it is first necessary to recall that pressure is defined as the normal

force per unit area and that the sum of the forces in each of three mutually
perpendicular directions must each be zero if a body is in equilibrium. Now
consider a small prism of fluid at a given point on a given horizontal plane, as
shown in Figure 2.6. Assume that the pressures p_1, p_2, and p_3 act normal to the

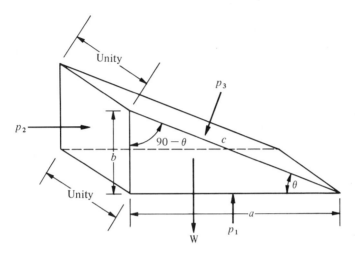

Figure 2.6. *Equilibrium in a fluid.*

respective faces as shown, that W is the weight of the prism, and that the depth
of the prism will be taken to be unity. Since the problem posed concerns a
horizontal plane, it will be unnecessary to consider forces perpendicular to the
prism at the end faces (they do not have components in the plane under consideration).

From the condition that the sum of the forces in the horizontal direction
must be zero,

$$p_2 b - p_3 c \cos(90 - \theta) = 0 \tag{2.9}$$

and from the condition that the sum of the forces in the vertical direction must
be zero,

$$p_1 a - p_3 c \cos \theta - W = 0 \tag{2.10}$$

and the geometry of the prism

$$a = c \cos \theta \tag{2.11a}$$

$$b = c[\cos(90 - \theta)] \tag{2.11b}$$

Substituting (2.11b) into (2.9) yields

$$p_2 = p_3 \qquad (2.12a)$$

and substituting (2.11a) into (2.10) yields

$$p_1 c \cos \theta - p_1 c \cos \theta = W \qquad (2.12b)$$

As the prism is made smaller until it becomes infinitesimal in size, W becomes zero and we have

$$p_1 = p_2 = p_3 \qquad (2.13)$$

From equation (2.13) we conclude that the pressure at a point is the same in every direction. Therefore the pressure exerted by a column of fluid on a given plane is independent of the shape of the column; it is only necessary to correctly apply equation (2.2) or (2.3) to obtain a solution.

Thus far the pressure within a liquid has been considered. The next logical question that arises is whether it is possible for a surface in contact with air and at rest to be other than horizontal. To answer this problem the same fundamental considerations will be invoked as before. First let us assume that a non-horizontal surface exists, as shown in Figure 2.7a. A free-body diagram of the

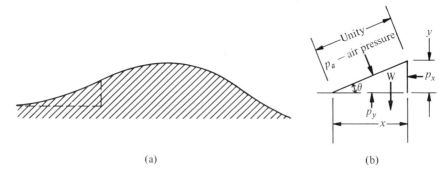

|(a)|(b)|

Figure 2.7. *Assumed conditions: nonhorizontal surface.*

liquid at the surface of the nonhorizontal portion is shown in Figure 2.7b. The liquid will be taken to have a unit width into the page. Also, we shall assume that there are no surface forces due to winds or other effects. Referring to Figure 2.7b and writing the required equilibrium equations,

$$p_y x - W - p_a \cos \theta = 0 \qquad (2.14a)$$

$$p_x y = p_a \sin \theta \qquad (2.14b)$$

However, since we have a surface in contact with the atmosphere, $p_y = p_a$, and either W must be zero or y must be zero if equilibrium is to exist. But y and W are both physical quantities that exist due to the assumption that a non-horizontal surface exists. Therefore, to satisfy the conditions of equilibrium it must be concluded that a static fluid must have a horizontal surface. Actually this conclusion and the conclusion that the pressure on a horizontal plane is the same at all points should be qualified. It would be more correct to state that an equipotential plane is one on which the pressure is the same at all points and that the free surface of a liquid in a container will conform to an equipotential plane. For the earth these planes will follow the earth's curvature. For most cases of engineering interest we may assume that a free surface (one in contact with the atmosphere) is a horizontal surface.

2.3 Pressure Measurement—Manometry[1]

Since pressure is a property of a system, its measurement is both desirable and necessary. In addition, as will be seen in later chapters, a knowledge of the pressure in various portions of a system is necessary to the determination of the flow and the thermodynamic state of a system. The simplest and most widely used mechanical pressure gage is the Bourdon gage. As shown in Figure 2.8, this gage consists simply of a bent tube closed on one end with a mechanism to indicate the movement of the closed end. The tube is usually noncircular in cross section and pressure is applied at the open end. When a pressure in excess of atmospheric pressure is applied to the open end, the closed end will move as the tube tends to straighten out (similar to a New Year's Eve blowout noise-maker) and the reading of the gage is proportional to the displacement of the tube. This type of gage can readily be calibrated and is rugged and easy to use. It is equally capable of measuring pressures above and below local atmospheric pressure. The student should note that the device illustrated measures gage pressure only. It is possible to evacuate the case of the gage and thereby create a gage capable of measuring absolute pressures.

If the gage is connected at an elevation different from the elevation where the pressure is to be measured, it is necessary to make a correction to the reading of the gage. Referring to Figure 2.8, it will be seen that if the gage is mounted above a pipe, the gage will read *low* by an amount equal to γh. The specific weight γ should be obtained at the temperature of the fluid in the connecting line to the gage since this may be different from the temperature in the pipe. For gages this correction is usually negligible.

[1]The student is referred to Appendix B for a more complete discussion of pressure measurement.

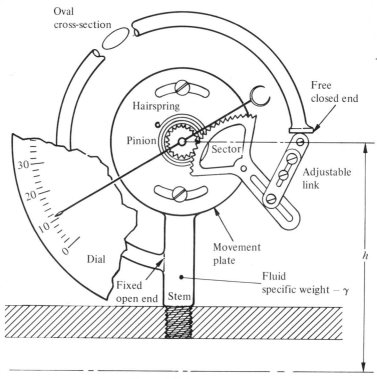

Figure 2.8. Bourdon gage.

We have already shown that the pressure at the base of a column of liquid is simply a function of the height of the column and the specific weight of the liquid. Therefore the height of a column of liquid of known specific weight can be and is used to measure pressure and pressure differences. Instruments that utilize this principle are known as manometers and the study of these pressure-measuring devices is known as manometry. By properly arranging a manometer and selecting the fluid judiciously it is possible to measure extremely small pressures, very large pressures, and pressure differences. A simple manometer is shown in Figure 2.9, where the right arm is exposed to the atmosphere while the left arm is connected to the unknown pressure. As shown, the fluid is depressed in the left arm and raised in the right arm until no unbalanced pressure forces remain. It has already been demonstrated that the pressure at a given level in either arm must be the same so that we can select any level as reference and write a relation for the pressure. Actually it is much easier and more convenient to select the interface between the manometer fluid and the unknown fluid as a common reference level. In Figure 2.9 the pressure at elevation AA is the same in both arms of the manometer. Starting with the open manometer arm (right) we have atmospheric pressure p_a acting on the fluid. As one proceeds down the arm the pressure increases until we arrive at level AA, where the pressure is $p_a + \gamma h$. This must equal the unknown pressure on the connected arm p_u. Therefore,

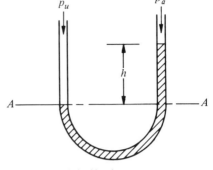

$$p_u = p_a + \gamma h \qquad (2.15a)$$

Figure 2.9. U-tube manometer.

or

$$p_u - p_a = \gamma h \qquad (2.15b)$$

ILLUSTRATIVE PROBLEM 2.3. A gas main, 10 in. in diameter, has water manometer gages provided at point A and point B, as shown in Figure 2.10. Point

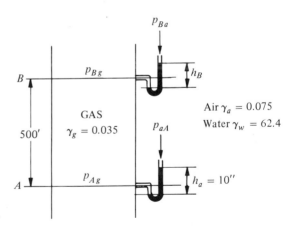

Figure 2.10. Illustrative problem 2.3.

B is 500 ft higher than point A. If the height of the water column at A is 10 in., what would be the height of the water column at b? (Assume no flow in pipe and a specific weight of gas $= 0.035 \text{ lb/ft}^3$; the specific weight of air is 0.075 lb/ft^3.)

Solution. In this problem the air and gas densities will be assumed to be constant. Also notice that the atmospheric pressure at B differs from (is less than) the atmospheric pressure at A. Let p_{Bg} be the pressure of gas at B, p_{Ba}

the pressure of air at B, p_{Ag} the pressure of gas at A, and p_{Aa} the pressure of air at A. (Refer to Figure 2.10.) The specific weight of water (γ_w) is 62.4 lb/ft³. Writing an equilibrium equation at B,

$$p_{Ba} + \gamma_w h_B = p_{Bg} \tag{2.16a}$$

Writing an equilibrium equation at A,

$$p_{Aa} + \gamma_w h_a = p_{Ag} \tag{2.16b}$$

But

$$p_{Ba} + 500\gamma_a = p_{Aa} \tag{2.16c}$$

and

$$p_{Bg} + 500\gamma_g = p_{Ag} \tag{2.16d}$$

Equations (2.16a) through (2.16d) completely determine the problem. Inserting the data of the problem and solving yields $h_B = 13.85$ in. An alternative way of looking at this problem is to say the gas pressure at B is less than that at A by an amount equal to 500 ft multiplied by γ_g. If the air pressure were constant, $\gamma_w h_B = \gamma_w h_a - 500\gamma_g$. But the air pressure at B is less than that at A by an amount equal to $\gamma_a \times 500$. Thus

$$\gamma_w h_B = \gamma_w h_a - \gamma_g(500) + 500\gamma_a \tag{2.16e}$$

or

$$\gamma_w h_B = \gamma_w h_a + 500(\gamma_a - \gamma_g) \tag{2.16f}$$

Solving equation (2.16f) directly yields $h_B = 13.85$ in.

The student will note that the height between B and the left leg of the top manometer has been neglected. Since this represents a gas column of inches of gas and is initially unknown, it has been neglected in the interest of simplicity. Neglecting this gas leg in each of the manometers introduces a negligible error into the final solution of this problem.

In the simple U-tube manometer it was assumed that the pressure was acting on the manometer fluid without considering how this came about. Let us consider the case of water (or other fluid) flowing in a pipe and a different fluid (say mercury or oil) being used in the manometer. This situation is shown in Figure 2.11. The pressure at level A is found from the equilibrium expression as follows:

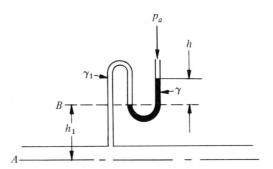

Figure 2.11. Manometer.

$$p_A + \gamma h + \gamma_1 h_1 = p_a \qquad (2.17)$$

If gage pressure is desired,

$$p_A = \gamma h + \gamma_1 h \qquad (2.18)$$

In equations (2.17) and (2.18) it is assumed that γ_1 is constant and the same in the manometer leg and the pipe. If this is not the case (due, say, to heat transfer), then a correction must be made.

ILLUSTRATIVE PROBLEM 2.4. Water ($\gamma_1 = 62.4 \text{ lb/ft}^3$) flows in a pipe. If a manometer connected as in Figure 2.11 is used with mercury as the fluid ($\gamma = 850 \text{ lb/ft}^3$) and h is 5 in. when the level B is 15 in. above the center line of the pipe, what is the pressure in the pipe?

Solution. Refer to Figure 2.11. From equation (2.18),

$$p_A = \gamma h + \gamma_1 h_1$$

$$\therefore p_A = 850 \times \frac{5}{12} + \frac{15}{12} \times 62.4 = 432.2 \text{ psfg}$$

or

$$p_A = \frac{432.2}{144} = 3 \text{ psig}$$

or

$$p_A = 14.7 + 3 = 17.7 \text{ psia}$$

To achieve greater accuracy and sensitivity in manometers, several different arrangements have been used. Perhaps the simplest of these is the inclined manometer. Consider a relatively large reservoir of liquid connected to a small bore tube that makes an angle θ with the horizontal. The pressure or pressure differential to be measured is connected to the large reservoir, while the inclined tube is open ended. Schematically this is shown in Figure 2.12. The unknown pressure p_u is given by,

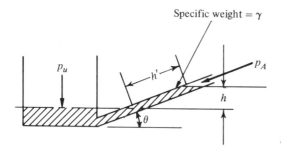

Specific weight $= \gamma$

Figure 2.12. *Inclined manometer.*

$$p_u = p_A + \gamma h \tag{2.19a}$$

or

$$p_u - p_A = \gamma h \tag{2.19b}$$

However, h is $h' \sin \theta$. Thus

$$p_u - p_A = \gamma h' \sin \theta \tag{2.20}$$

Since θ is fixed, a scale placed along the tube can be calibrated to read directly in units of h of a fluid. Usually, this is done by reading directly inches of water for $p_u - p_A$.

Another method of achieving greater sensitivity and accuracy is to use a manometer with more than one fluid. The manometer shown in Figure 2.11 can be used for this purpose by properly selecting the manometer fluid. The arrangement shown in Figure 2.13 can also be used in this manner. Starting at level A,

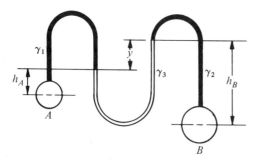

Figure 2.13. *Three-fluid manometer.*

$$p_A - h_A\gamma_1 - y\gamma_3 + h_B\gamma_2 = p_B \qquad (2.21a)$$

or

$$p_A - p_B = (-h_B\gamma_2 + h_A\gamma_1) + y\gamma_3 \qquad (2.21b)$$

In the usual case the manometer is connected to different positions on the same pipe and γ_1 can be taken to be equal to γ_2. Also, $h_B - h_A = y$. Thus

$$p_A - p_B \simeq y(\gamma_3 - \gamma_1) \qquad \text{or} \qquad y(\gamma_3 - \gamma_2) \qquad (2.22)$$

For small differences in $p_A - p_B$ it is apparent that the manometer fluid (γ_3) should have a specific weight very nearly equal to the specific weight of the fluid in the pipes. For large pressure differences one can use a heavy fluid such as mercury to increase $\gamma_3 - \gamma_2$ and reduce the manometer reading. The student should note that it was assumed that A and B were at the same level. If this is not the case, equations (2.21a) through (2.22b) should be corrected for the difference in level.

Another way to increase the sensitivity of a manometer is shown in Figure 2.14. The ends of this manometer consist of larger (and equal) areas contain-

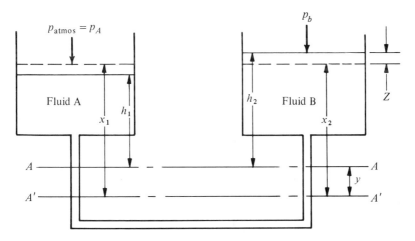

Figure 2.14. Sensitive U-tube manometer.

ing two fluids that will not mix. Before any pressure is applied, the common surface of the two liquids is at AA and the heights of the two liquids above this level are h_1 and h_2. Since the fluids must be in equilibrium,

$$h_1\gamma_1 = h_2\gamma_2 \qquad (2.23)$$

where γ_1 and γ_2 are the specific weights of the fluids. A pressure is applied to the right arm of the tube, which causes the fluids to be displaced, and the common surface descends an amount y to the new level $A'A'$. The levels of the two liquids above the common surface are now x_1 and x_2, and the applied pressure p_b is given by,

$$p_A + x_1\gamma_1 = x_2\gamma_2 + p_b \tag{2.24a}$$

or

$$p_b - p_A = x_1\gamma_1 - x_2\gamma_2 \tag{2.24b}$$

Since mass must be conserved, the mass of fluid displaced in one end must equal the mass of fluid displaced in the other end, or

$$AZ = ay \tag{2.25}$$

where A is the area of cross section of large ends—both taken to be equal—and a the area of cross section of the tube. From geometry,

$$\begin{aligned} X_1 &= h_1 + Z + y \\ X_2 &= h_2 - Z + y \end{aligned} \tag{2.26}$$

Substituting these in equation (2.24) gives

$$p_b - p_A = \gamma_1(h_1 + Z + y) - \gamma_2(h_2 + Z + y)$$

and

$$p_b - p_A = \gamma_1(Z + y) - \gamma_2(y - Z)$$

but

$$Z = \frac{a}{A} y$$

Thus

$$p_b - p_A = y\left[\gamma_1\left(1 + \frac{a}{A}\right) - \gamma_2\left(1 - \frac{a}{A}\right)\right] = y\left[(\gamma_1 - \gamma_2) + (\gamma_1 + \gamma_2)\frac{a}{A}\right] \tag{2.27}$$

From equation (2.27) we conclude that for greatest sensitivity γ_1 and γ_2 should be as close as possible and the area ratio a/A should be as small as possible.

ILLUSTRATIVE PROBLEM 2.5. A three-fluid manometer such as the one shown in Figure 2.13 is to be used to measure large pressure differences. Assuming that the same liquid is in the legs with $\gamma = 62.4\ \text{lb/ft}^3$ (water) and that the manometer fluid is mercury ($\gamma_3 = 850\ \text{lb/ft}^3$), determine the pressure difference between A and B if y is equal to 1 ft and A and B are at the same level.

Solution. From equation (2.22), which is applicable to this situation,

$$p_A - p_B \simeq y(\gamma_3 - \gamma_1) = 1(850 - 62.4) = 787.6\ \text{psfg}$$

and

$$p_A - p_B = \frac{787.6}{144} = 5.47\ \text{psig}$$

ILLUSTRATIVE PROBLEM 2.6. If in illustrative problem 2.5, A is 6 in. higher than B, determine $p_A - p_B$.

Solution. Refer to Figure 2.13. Starting from A,

$$p_A - h_A \gamma_1 - y\gamma_3 + h_B \gamma_2 = p_B$$

Assuming $\gamma_1 = \gamma_2 = \gamma$,

$$p_A - p_B = \gamma(h_A - h_B) + y\gamma_3$$

From the statement of the problem,

$$h_B - \left(\frac{1}{2} + h_A\right) = y$$

or

$$h_B - h_A = y + \frac{1}{2}$$

and thus

$$p_A - p_B = \gamma\left(-y - \frac{1}{2}\right) + y(\gamma_3) = y(\gamma_3 - \gamma) - \frac{1}{2}\gamma$$

This result could almost have been anticipated by observation; that is, elevating one leg by $\frac{1}{2}$ ft is equivalent to decreasing the pressure difference by $\frac{1}{2}$ ft of fluid in the leg. Therefore

$$p_A - p_B = 1(850 - 62.4) - \frac{1}{2}(62.4) = 756.4\ \text{psfg}$$

or

$$p_A - p_B = \frac{756.4}{144} = 5.25 \text{ psig}$$

For many applications it is desired to read the pressure in a system remotely or to obtain an electrical signal that is proportional to the pressure. The term *transducer* will be used to describe any device that will convert a pressure signal into an electrical signal. In general, two classes of instruments are available for pressure measurement.

The first class of transducer is the piezoelectric transducer, which is based upon the principle that a crystal under stress can generate a voltage. Quartz crystals are commonly used and depending on the cut and mounting of the crystal, the transducer will either respond directly to pressure or to the displacement of the element.

The second class of transducer utilizes the change in the electric circuit parameters rather than the generation of a voltage. This change in circuit parameter (resistance, inductance, or capacitance) is then observed or amplified for the desired measurement. One of the common transducers of this type is the strain gage, which depends on the change in resistance of a fine wire when the length of the wire changes. The strain gage pressure transducer is a common industrial tool. In addition, the movement of an element can be used to vary either the inductance or capacitance of a unit and thus to produce electrical changes proportional to the applied pressure. Capacitive transducers have been used in microphones to respond to the minute pressure variation caused by sound waves and also to record explosive blast pressures.

Transducers can be used to measure either gage or absolute pressures. It is imperative that each transducer be capable of being periodically calibrated to assure that the output is related to the actual pressure to be measured. Direct calibration is sometimes very difficult or impossible, and it then becomes necessary to resort to calibrations against some from of secondary standard, such as a Bourdon gage.

2.4 Forces on Plane–Submerged Surfaces

It was seen in Chapter 1 that the shear stress in a fluid is proportional to the velocity difference between the shearing surfaces and inversely proportional to the distance separating the surfaces. If a fluid is at rest, the shear stress must be zero, which means that no tangential forces can exist in the fluid. The only forces that can exist are forces that are normal to surfaces in contact with the fluid. If a plane is completely immersed in a fluid it will experience a normal force on both sides of the plane, which will tend to compress the plane. However, when a plane is placed in a fluid so that only one side is subjected to the fluid pressure, it will experience a net unbalanced force. Since pressure varies with depth, the pressure on a nonhorizontal plane will in general be variable.

First let us consider the special case of a vertical plane and subdivide this topic into (1) a vertical plane just touching a free surface, (2) a completely submerged vertical plane, and (3) a vertical plane with different liquid levels on each side.

For the vertical plane just touching a free surface, case 1, the pressure on one side will vary linearly from zero to γh due to the column of fluid on this side. This is shown schematically in Figure 2.15a, where the pressure is indicated

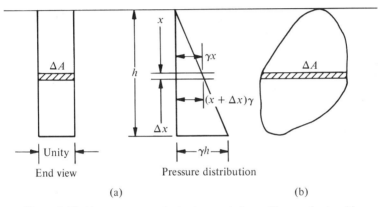

Figure 2.15. Pressure on a vertical submerged plane with one edge touching a free surface.

to vary linearly, giving rise to a triangular pressure distribution. If the plane is a unit width into the paper, the force exerted on the area, $(\Delta x)(1)$, is the average pressure on this area multiplied by the area. The average pressure is

$$p_{\text{avg}} = \frac{\gamma x + \gamma(x + \Delta x)}{2} = \gamma x + \frac{\gamma \Delta x}{2} \tag{2.28}$$

The force on Δx is thus

$$F_x = \gamma x(\Delta x) + \frac{\gamma(\Delta x)(\Delta x)}{2} \tag{2.29}$$

Since Δx is small, the term $(\Delta x)^2$ is indeed very small and may be neglected when compared to terms containing only Δx. Therefore, the second term in equation (2.29) can be neglected when compared with the first term. Thus

$$F_x = \gamma x \, \Delta x \qquad (2.30a)$$

or since $(\Delta x)(1) = \Delta A$,

$$F_x = \gamma x \, \Delta A \qquad (2.30b)$$

It is necessary to sum each of the F_x terms over all values of x in order to evaluate the total force on the plane. For the special case where the width of the plane is constant it will be immediately seen that A, the sum of all the ΔA's, is not a function of x and one need only use the average value of x to obtain the total force on the plane. Thus, since x varies from zero to h,

$$F_x = \frac{\gamma h A}{2} \qquad (2.31)$$

Note that $h/2$ represents the center of gravity (or centroid) of a rectangle and that the average pressure on the area is just $\gamma h/2$ or $\gamma \bar{h}$, where $\bar{h}$ is the location of the center of gravity from the liquid surface.

Let us now consider a plane whose width is not constant into the paper but varies in some manner with liquid depth. This situation is shown in Figure 2.15b. The pressure on the small area ΔA can be written as

$$p = \frac{F_x}{A} = \gamma x \, \Delta A \qquad (2.32)$$

and the pressure over the entire plate as

$$p = \frac{F}{A} = \sum \gamma x \, \Delta A \qquad (2.33a)$$

or

$$p = \sum \frac{\gamma x \, \Delta A}{A} \qquad (2.33b)$$

where $\sum$ denotes the sum of all such quantities.

The student will note from elementary physics that $\sum (x \, \Delta A / A)$ is the definition of the location of the center of gravity or centroid of any area. Thus,

$$p = \gamma \bar{h} \qquad (2.34)$$

where $\bar{h}$ is the location of the center of gravity of the area from the surface. Thus from equation (2.34), we see for a vertical submerged plane touching the surface of a fluid that the average pressure on the plane is simply the product of the specific weight of the fluid multiplied by the depth of the center of gravity from the surface of the fluid. The total force is the total area multiplied by this average pressure. Therefore,

$$F_{\text{total}} = \gamma \bar{h} A \qquad (2.35)$$

The second case of a vertical plane, case 2, is one that is completely submerged. Figure 2.16 shows any shaped vertical plane whose top edge is located

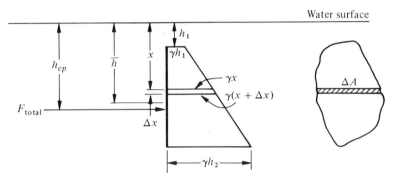

Figure 2.16. *Submerged vertical surface.*

some distance below the free surface of the liquid. Again the pressure on one side will vary linearly with the depth of fluid, and the plot of pressure on one side of the plane against the distance below the top edge of the plate is a trapezoid whose value of pressure on the top edge is γh_1 and on the bottom edge is γh_2. The force on the area ΔA is the average pressure multiplied by the area. Therefore,

$$\frac{F_x}{\Delta A} = \gamma x + \frac{\gamma \, \Delta x}{2} \qquad (2.36)$$

The force on ΔA is therefore

$$F_x = \left(\gamma x + \frac{\gamma \, \Delta x}{2}\right) \Delta A \qquad (2.37)$$

Carrying out the multiplication indicated by equation (2.37) and neglecting products of small numbers and adding all such items yields

$$F_x = \sum \gamma x \, \Delta A \tag{2.38}$$

Since pressure is force divided by area, the average pressure on the plate is

$$p = \sum \frac{\gamma x \, \Delta A}{A} \tag{2.39}$$

where A is the total plate area and $\sum$ is the sum of all such quantities over the plate. Again $\sum (x \, \Delta A / A)$ is simply the location of the center of gravity or centroid of the plate. Thus we have the result that for a vertical submerged plate the average pressure on a surface is simply the product of the specific weight of the fluid multiplied by the depth of the center of gravity below the surface of the fluid. Also, as before, the total force is the total area multiplied by the average pressure. This leads to the result that equation (2.35) is general for any vertical surface, namely $F_{\text{total}} = \gamma \bar{h} A$, where $\bar{h}$ is the location of the center of gravity of the plane below the surface of the fluid.

ILLUSTRATIVE PROBLEM 2.7. A vertical plate is used to dam a channel 2 ft wide. If the plate is $2\frac{1}{2}$ ft high and the water on one side of the plate is at a level $\frac{1}{2}$ ft below the top, what is the total force on the plate?

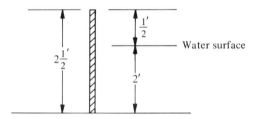

Solution. For this problem, the $2\frac{1}{2}$ ft dimension is not important. Only the portion of the plate that is submerged is important. Its center of gravity is 1 ft below the free liquid surface and the area in contact with the liquid is 2×2 ft or 4 ft². The total force on the plate is therefore

$$F_{\text{total}} = \gamma \bar{h} A = 62.4 \times 1 \times 4 = 249.6 \, \text{lb}$$

As the last case of a vertical surface subjected to hydrostatic forces, case 3, consider the plane subjected to different liquid levels (but the same liquid) on both sides. It is possible to obtain the net force acting on this plane by considering each side separately and vectorially summing the forces. By considering Figure 2.17 a simple expression for the net force for this case can readily be developed. The pressure distribution for the left side is a triangle extending from zero at the upper free surface to γh_1 at the lower edge of the plate. On

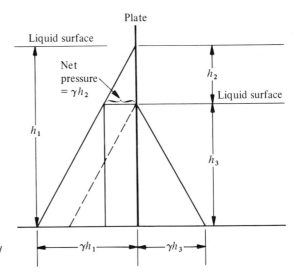

Figure 2.17. *General submerged vertical surface.*

the right-hand side the pressure distribution is also a triangle extending from zero at its free surface to γh_3 at the base of the plate. By superposing these two diagrams as shown by the dashed line on the figure, we reach the conclusion that from h_2 down the pressure is constant and equal to γh_2 on the plate. The total force on the plate is therefore given by

$$F_{\text{total}} = \gamma h_2 A_1 + \gamma \frac{h_2}{2} A_2 \qquad (2.40)$$

where A_1 is the plate area below the right liquid surface and A_2 the area above the right liquid surface.

Illustrative Problem 2.8. A vertical plate extends from the surface of water on one side to a depth of 10 ft. On the other side there is water to within 2 ft of the top. If the plate is 5 ft wide, what is the net force on the plate?

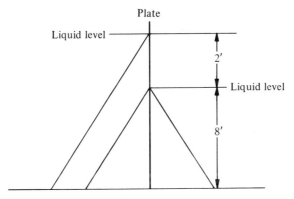

Solution. This problem can be solved by considering each side separately. For the left side,

$$F_{total} = \gamma \bar{h} A = 62.4 \times 5 \times 10 \times 5 = 15,600 \text{ lb}$$

On the right side,

$$F_{total} = \gamma \bar{h} A = 62.4 \times 4 \times 8 \times 5 = 10,000 \text{ lb}$$

The net force $= 15,600 - 10,000 = 5600$ lb.

An alternative solution is obtained by considering the superposed pressure distribution as shown in the figure. For the upper 2 ft, we have

$$F_{total} = 1 \times 2 \times 5 \times 62.4 = 624 \text{ lb}$$

For the lower 8 ft the pressure is uniform and the force is therefore given by equation (2.40) as

$$\gamma \times 2 \times A = 4980 \text{ lb}$$

The total force is given as

$$F_{total} = 624 + 4980 = 5604 \text{ lb}$$

Both methods yield the same result and it is left to the student to use the one of his choice, or, better yet, to do this type of problem by both methods to obtain an independent check on his solution.

Vertical and horizontal surfaces represent special cases of submerged planes, and, in general, the submerged plane can make any angle with the free surface of the liquid. Consider the plane shown in Figure 2.18, which makes an angle θ with the liquid surface. Also recall that it was proved earlier in this chapter that the pressure intensity on any horizontal plane must be the same anywhere on the plane. Thus the pressure at the level x is the same regardless of the orientation of the surface. However, this pressure acts normal to the surface, which in this case is normal to Δy and over the area ΔA. Using the same procedure as before leads us to the solution that

$$p = \gamma \bar{L} \sin \theta \qquad (2.41)$$

and since $L \sin \theta = \bar{h}$,

$$p = \gamma \bar{h} \qquad (2.42)$$

and

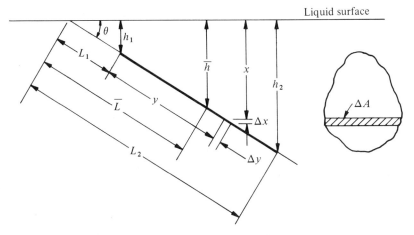

Figure 2.18. *Inclined submerged plane.*

$$F_{\text{total}} = \gamma \bar{h} A \tag{2.43}$$

It can now be stated as a general result that the total force acting on a submerged plane area is the product of the specific weight of the liquid multiplied by the area and multiplied by the depth of the center of gravity (or centroid) of the area from the liquid surface. This conclusion does not depend on the angle of the plane area.

ILLUSTRATIVE PROBLEM 2.9. A valve (plate) is placed in a channel to control the flow. Assume that the valve consists of a plane that makes an angle of 30 deg with the free surface. If the channel is 5 × 5 ft in cross section and is flowing full, what is the total force on the valve after the flow is cut off? Assume $\gamma = 62.4 \, \text{lb/ft}^3$.

Solution. The area of the valve is 5 × 5/sin 30 = 50 ft². The centroid of

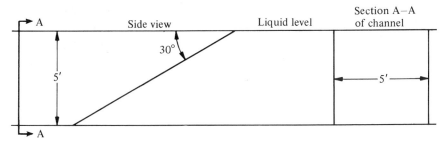

the plate is located $2\frac{1}{2}$ ft below the surface. The total force is therefore

$$2\tfrac{1}{2} \times 50 \times 62.4 = 7800 \, \text{lb}$$

2.5 Location of the Center of Pressure

In Section 2.4 the discussion was concerned with the magnitude of the force on a plane area. Since force is a vector quantity, a force is completely described by knowing its magnitude, direction, and point of application. Thus far we have determined the magnitude and direction but not the point of application of these vectors (forces). To determine the point of application of each of the forces, let us once more consider the case of a fully submerged vertical plane such as is shown in Figure 2.16. Since the area is in equilibrium, the sum of the moments about any point must be zero. Therefore, by referring to Figure 2.16 and using the notation h_{cp} for the vertical location of the point of application of the resultant force,

$$(F_{\text{total}})h_{cp} = \sum \gamma x \, \Delta A(x) \tag{2.44a}$$

and

$$h_{cp} = \sum \frac{\gamma x^2 A}{F_{\text{total}}} \tag{2.44b}$$

But

$$F_{\text{total}} = \gamma \bar{h} A.$$

Therefore

$$h_{cp} = \sum \frac{x^2 \, \Delta A}{\bar{h} A} \tag{2.45}$$

The summation $\sum x^2 \, \Delta A$ term in equation (2.45) is the second moment of area, or more conventionally it is called the moment of inertia (I) of the area A about the axis at the surface of the liquid. Actually, it is more convenient to express equation (2.45) in terms of the moment of inertia about the center of gravity of the area. To do this we utilize the transfer theorem of mechanics, and inserting the results into equation (2.45), we obtain

$$h_{cp} = \frac{\bar{I}}{\bar{h} A} + \bar{h} \tag{2.46}$$

where $\bar{h}$ is the location of the center of gravity and $\bar{I}$ is the moment of inertia of the plane area about a centroidal axis. The location of the resultant force is known as the center of pressure. Table 2.2 gives the properties of some selected areas.

Table 2.2 Properties of an Area

Section	Area	$\bar{h}$ location of center of gravity	$\bar{I}$ moment of inertia about the center of gravity
Square	A^2	$\frac{1}{2} A$	$\bar{I} = \frac{1}{12} A^4$
Rectangle	bd	$\frac{1}{2} d$	$\frac{1}{12} bd^3$
Trapezoid	$\frac{1}{2}(B + b) d$	$d \left\{ \dfrac{2B + b}{3(B + b)} \right\}$	$\dfrac{d^3(B^2 + 4Bb + b^2)}{36(B + b)}$
Triangle	$\frac{1}{2} bd$	$\frac{2}{3} d$	$\dfrac{bd^3}{36}$
Circle	πR^2	R	$\dfrac{\pi R^4}{4}$
Ellipse	$\dfrac{\pi bd}{4}$	$\frac{1}{2} d$	$\dfrac{\pi bd^3}{64}$
Quarter Circle — Also * for Semicircle	$\dfrac{\pi R^2}{4}$ $* \dfrac{\pi R^2}{4}$	$0.4244\,R$ $*0.4244\,R$	$\dfrac{\pi R^4}{8}$ $* \dfrac{\pi R^4}{8}$

The student should note that it is always possible to evaluate either $\bar{h}$ or $\bar{I}$ numerically by subdividing the area in question into many small parts and performing the indicated mathematical operations and summation. Even a relatively coarse subdivision quickly leads to a good numerical approximation of the exact solution.

ILLUSTRATIVE PROBLEM 2.10. A sluice gate is 6 ft square. It is hinged at the top and held by two horizontal pins 1 ft from the bottom of the gate. What force must each of these pins resist if the water level is at the top of the gate?

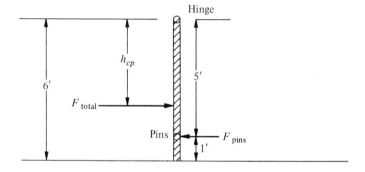

Solution. For the gate to be in equilibrium under the action of the forces shown, it is necessary that the moment of these forces about the hinge be zero. The hydrostatic force

$$F_{\text{total}} = \frac{6}{2} \times 6 \times 6 \times 62.4 = 6740 \text{ lb}$$

The location of h_{cp} is

$$\frac{\bar{I}}{\bar{h}A} + \bar{h} = \frac{\dfrac{bd^3}{12}}{\dfrac{d}{2(d)(b)}} + \frac{d}{2} = \frac{d}{6} + \frac{d}{2} = 4 \text{ ft}$$

The moment of these forces about the hinge is therefore

$$(F_{\text{total}})4 = (F_{\text{pins}})5$$

$$\therefore F_{\text{pins}} = \frac{4}{5}(6740) = 5392 \text{ lb}$$

and for one pin

$$F_{\text{pin}} = 2696 \text{ lb}$$

ILLUSTRATIVE PROBLEM 2.11. Flashboards are installed on top of a dam and are supported by pipes spaced 4 ft apart along the dam. If the water is flowing over the flashboard 1 ft deep as shown in the accompanying figure, what is the bending moment produced at the bottom of the flashboard on one pipe?

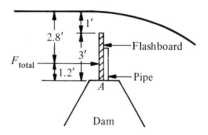

Solution. To solve this problem let us select a repeating section of flashboard 4 ft wide. The total force on the flashboard is

$$F_{\text{total}} = 62.4 \times 4 \times 3(1 + 1.5) = 1870 \text{ lb}$$

The location of h_{cp} is

$$h_{cp} = \frac{\dfrac{bd^3}{12}}{\bar{h}A} + \bar{h}$$

and

$$h_{cp} = \frac{4 \times 3^3}{(1 + 1.5)(4 \times 3)} + 2.5 = 2.8 \text{ ft from the water surface}$$

Taking moments about A, $1.2 \times 1870 = 2240$ ft lb. Since per repeating section there is the equivalent of one pipe, this is the moment at the bottom of a pipe.

For a nonvertical plane, as is shown in Figure 2.18, we can also evaluate the location of the center of pressure in a manner similar to that used for the vertical plane. When this is done using calculus, it can be demonstrated that

$$h_{cp} = \left(\frac{\bar{I}}{\bar{h}A} + \bar{h} \right) \tag{2.47}$$

Since equations (2.46) and (2.47) are identical, it follows that the vertical distance from a liquid surface to the center of pressure is the same regardless of the angle of inclination as long as the angle is not zero. The center of pressure is always below the center of gravity.

ILLUSTRATIVE PROBLEM 2.12. Determine the center of pressure for the valve in illustrative problem 2.9.

Solution

$$h_{cp} = \frac{\bar{I}}{\bar{h}A} + \bar{h}$$

For this problem all dimensions will be considered *along* the plane:

$$\bar{I} = \frac{bd^3}{12} = \frac{5 \times \left[\dfrac{5}{\sin 30}\right]^3}{12} = \frac{5 \times (10)^3}{12} = \frac{5000}{12}$$

$$\bar{h} = \frac{2.5}{\sin 30} = 5 \text{ ft}$$

$$A = 5 \times 10 = 50 \text{ ft}$$

$$\therefore h_{cp} \text{ along plane} = \frac{\frac{5000}{12}}{5 \times 50} + 5 = 1.67 + 5 = 6.67 \text{ ft}$$

For the vertical height h_{cp},

$$h_{cp} = 6.67 \sin 30 = 3.34 \text{ ft vertically below the surface}$$

2.6 Buoyancy of Submerged Bodies

If a body is weighed in a vacuum and then is weighed in a fluid it will be found to have a different weight in both cases. This is due to the buoyant action of the fluid displaced. To evaluate this effect let us consider the body shown in Figure 2.19. This body will be considered to be floating in a sub-

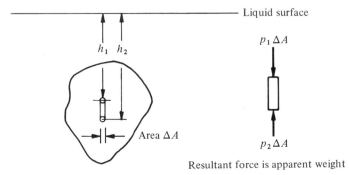

Resultant force is apparent weight

Figure 2.19. *Submerged body.*

merged position and therefore it will be in equilibrium in this position. Obviously this is a problem in hydrostatics that can be resolved by applying the principles already developed in this chapter by considering the small volume shown in Figure 2.19 and summing all such volumes.

From the free-body diagram shown we need consider only the vertical forces since all horizontal forces on the body are equal and opposite and give no net horizontal force. The vertical forces are the weight of the body and the forces on each end of the body, with the resultant of these forces equal to the sum of these three vectors. Thus

$$\text{Apparent weight (resultant)} = W + p_1\,\Delta A - p_2\,\Delta A \qquad (2.48a)$$

$$= \gamma_s\,\Delta V + \gamma h_1\,\Delta A - \gamma h_2\,\Delta A \qquad (2.48b)$$

where γ_s is the specific weight of the solid and γ the specific weight of the liquid. The volume of the element is given as ΔV and equals $(h_2 - h_1)\,\Delta A$. Therefore,

$$\text{Apparent weight} = \gamma_s\,\Delta V - \gamma\,\Delta V = \Delta V(\gamma_s - \gamma) \qquad (2.49)$$

Notice that the apparent weight equals the weight of the body in vacuum less the weight of the fluid displaced. If all of the small volumes in the body are summed the total volume of the body will be obtained. Thus

$$\text{Apparent weight of the submerged body} = V(\gamma_s - \gamma) \qquad (2.50)$$

Stated in words, a submerged body is buoyed up by a force equal to the weight of the displaced fluid, and the buoyant force must also act through the center of gravity of the displaced fluid for the body to be in equilibrium. This is known as Archimedes' principle.

ILLUSTRATIVE PROBLEM 2.13. A large stone weighs 100 lb in air, and when the stone is immersed in water it weighs 60 lb. Calculate the volume of the stone in cubic feet and its specific weight.

Solution. Refer to Figure 2.19 and neglect the buoyant effect in air. Thus the buoyant effect in water is the difference between the weight in air and the weight in water or $100 - 60 = 40$ lb. This buoyant effect equals the weight of water displaced or the volume of displaced water multiplied by the specific weight of water:

$$\gamma V = 40$$

$$V = \frac{40}{62.4} = 0.64 \text{ ft}^3$$

The specific weight is simply weight divided by volume:

$$\gamma_s = \frac{W}{V} = \frac{100}{0.64} = 156 \text{ lb/ft}^3$$

2.7 Forces on Curved Surfaces

Thus far we have considered forces acting on flat plane surfaces. Very often a curved surface is used in engineering work, and it is necessary to determine the forces acting on such surfaces. In general the problem of a resultant force on an irregular surface reduces to the solution of a nonparallel, noncoplanar force system. However, for most cases of interest it is found that the surface in question is symmetrical and that the horizontal component of the total force and the vertical component of the total force can be directly combined to yield a single force. It is therefore convenient to evaluate the horizontal and vertical forces on a curved submerged surface to obtain the resultant total force.

The student will again note that a force is a vector quantity. Since vectors cannot be added algebraically, the force on each unit of area must be broken into two mutually perpendicular components, and these components must be added. Since each of these unit areas makes a different angle with the coordinates, the vectors also make differing angles with the coordinates and this process is sometimes virtually impossible to do analytically.

The principles used in solving problems involving curved surfaces are best illustrated by discussing a specific problem. Consider the quarter circle shown in Figure 2.20. For the surface AE to be in equilibrium it is necessary that the

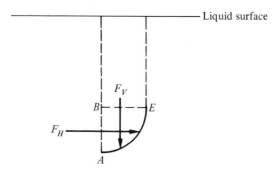

Figure 2.20. *Submerged vertical surface.*

reactant equal the sum of the forces acting on the surface. The sum of the horizontal forces must equal the force on the projected area AB. This follows, since the fluid cannot sustain shear forces. If this were not true, a vertical plane placed at AB would not be in equilibrium. Also, the location of F_H is the center of pressure of the projected area AB. For vertical equilibrium, the surface AE must support the weight of fluid above it. This weight is best considered to consist of the force on the horizontal projected area BE plus the weight of fluid contained by the curved volume ABE. Each of these vertical forces will act through their respective centers of gravity, which can be obtained from Table 2.1 for the various geometric shapes.

ILLUSTRATIVE PROBLEM 2.14. On the top of a dam a radial gate is mounted as shown in the accompanying figure with its top edge even with the water level. The gate extends down to become tangent to the top of the dam and has a radius of 12 ft and a length of 12 ft. What is the total water pressure against the gate?

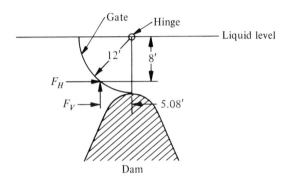

Solution. In this case the hydrostatic pressure acts on the outside of the curved surface, and there is a vertical buoyant force on the surface. This is equivalent to a negative (vertically upward) force of the same magnitude and location as if the gate were filled with fluid. Proceeding to calculate the vertical force, we simply have the weight of the displaced fluid in the quarter circle multiplied by the specific weight of the fluid;

$$F_V = \tfrac{1}{4}\pi R^2 L \gamma,$$

where L is the length of the gate,

$$\therefore F_V = \tfrac{1}{4}\pi(12)^2 \times 12 \times 62.4 = 84{,}600 \text{ lb}$$

The location of the center of gravity $\bar{h}$ is given from Table 2.1 as $0.4244R = 0.4244 \times 12 = 5.08$ ft. For the horizontal force, we simply apply the results determined previously in this chapter. Therefore,

$$F_H = 12 \times 12 \times 6 \times 62.4 = 53{,}900 \text{ lb}$$

$$h_{cp} = \frac{\bar{I}}{\bar{h}A} + \bar{h} = \frac{\dfrac{12(12)^3}{12}}{6 \times 12 \times 22} + 6 = 8 \text{ ft}$$

The total force is the resultant of F_V and F_H and equals $(84,600)^2 + (53,900)^2 = 100,000$ lb. This forces acts through the point 8 ft below the water level and 5.08 ft to the left of the vertical, as shown in the figure.

2.8 Stresses in Cylinders and Spheres

When a fluid is contained, it exerts forces on the walls of the container that stress the material of the container. As an extension of the principles discussed in this chapter let us consider the problem of evaluating the stresses in thin cylinders and spheres subjected to an internal pressure. To evaluate these stresses we shall assume that the ratio of the thickness to the diameter of the cylinder or sphere is small and that the stresses developed in these pressure vessels are uniform over the cross section of the vessel. By *thin* we shall mean that the ratio of thickness to diameter is 0.1 or less.

With these assumptions in mind, let us consider the cylinder shown in Figure 2.21. Due to the internal pressure there is a stress in the circumferential

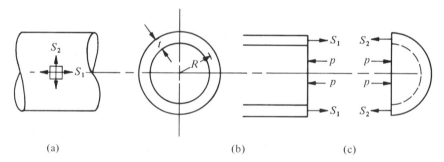

Figure 2.21. *Thin cylinder under internal pressure.*

direction (S_2) and a stress in the longitudinal direction (S_1), as shown in Figure 2.21a. If the pipe is cut perpendicular to the longitudinal axis (and far from the ends) the resultant free-body diagram will be as shown in Figure 2.21b. The total resisting force in the cylinder will be the stress S_1 multiplied by the area over which the stress acts. Thus the resisting force is $S_1(2\pi R)(t)$ since the material area is $2\pi Rt$. The applied load is due to the pressure in the tube acting over the tube area. The applied load is therefore $p\pi R^2$. For equilibrium these forces must be equal. Therefore,

$$S_1 2\pi Rt = p\pi R^2$$

or

$$S_1 = \frac{pR}{2t} \tag{2.51}$$

By considering a diametrical cut through the cylinder, as shown in Figure 2.21c, S_2 can be evaluated in an analogous manner. The area is $2tL$, where L is the pipe length. Acting on the pipe we have the internal force component in the horizontal direction, which is $p2RL$, where $2RL$ is the projected area of the tube. Thus,

$$2tLS_2 = p2RL$$

and

$$S_2 = \frac{pR}{t} \tag{2.52}$$

From equations (2.51) and (2.52) it can be seen that S_2 is twice S_1. The student should also note that these stresses were arrived at without having to involve the material of the cylinder. The strength of the material will determine the thickness of the cylinder required for a given internal pressure, and the interested student is referred to any standard text on strength of materials for further discussion of this subject.

Let us now consider a thin sphere subjected to an internal pressure. In this sphere S_1 will be equal to S_2 by symmetry. As can be seen from Figure 2.22,

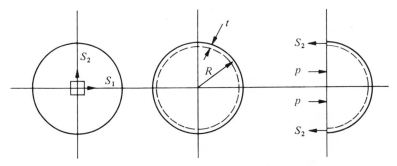

Figure 2.22. Thin sphere under internal pressure.

the stress S_2 acts over an area equal to $2\pi Rt$. The sum of the force components in the horizontal direction is once again just the product of the pressure multiplied by the projected area, that is, $p\pi R^2$. Therefore,

$$S_2 2\pi Rt = p\pi R^2$$

or

$$S_2 = S_1 = \frac{pR}{2t} \qquad (2.53)$$

Equation (2.53) yields the result that S_2 in a thin sphere is half of S_2 in a thin cylinder when both are subjected to the same internal pressure. The student is cautioned to consult a text on strength of materials or applicable codes before attempting to design cylinders and spheres for either internal or external pressure.

ILLUSTRATIVE PROBLEM 2.15. A 48-in.-diameter steel pipe 0.25 in. thick carries water under an internal pressure of 100 psi.
1. Compute the circumferential stress in the steel.
2. If the pressure is raised to 250 psi and the allowable unit stress in the steel is 15,000 psi, what thickness of steel is required?

Solution

1. The circumferential stress is S_2. Thus from equation (2.52),

$$S_2 = \frac{pR}{t} = \frac{100 \times 24}{0.25} = 9600 \text{ psi}$$

2.

$$t = \frac{pR}{S_2} = \frac{250 \times 24}{15,000} = 0.4 \text{ in.}$$

2.9 Closure

Fluid statics is simply an extension of the principles of statics that the student studied for solid bodies in his physics and mechanics courses. However, this is not to be taken to minimize this division of fluid mechanics since the applications of these principles occur frequently and in all fields of engineering. Some of these applications have been explicitly pointed out and studied in this chapter, but there are many others that are encountered, and even though we were, of necessity, limited to the number of items studied, mastery of these items will enable the student to readily analyze the novel situations that he may meet in his career.

In later chapters of this text we shall utilize many of the principles developed in our study of fluid statics. However, of necessity we shall require further information when dealing with fluids that are in motion relative to a given gravitational field. These items will be developed in each of the chapters that follow, as required, and we may consider this chapter on fluid statics as the first step toward the more complicated study of fluid dynamics.

REFERENCES

1. *Fundamentals of Hydro-and Aero-Mechanics* by L. Prandtl and O. G. Tietjens, McGraw-Hill Book Company, Inc., New York, 1934.

2. *Basic Fluid Mechanics* by J. L. Robinson, McGraw-Hill Book Company, Inc., New York, 1963.

3. *Fluid Mechanics For Engineers* by P. S. Barna, Butterworth & Co. (Publishers) Ltd., London, 1957.

4. *Fluid Mechanics* by V. L. Streeter, 3rd ed., McGraw-Hill Book Company, Inc., New York, 1962.

5. *Elementary Fluid Mechanics* by J. K. Vennard, John Wiley & Sons, Inc., New York, 1961.

6. *Elementary Theoretical Fluid Mechanics* by K. Brenkert, Jr., John Wiley & Sons, Inc., New York, 1960.

7. *Engineering Applications of Fluid Mechanics* by J. C. Hunsaker and B. G. Rightmire, McGraw-Hill Book Company, Inc., New York, 1947.

8. *Electronics in Engineering* by W. R. Hill, McGraw-Hill Book Company, Inc., New York, 1961.

9. "Pressure and Its Measurement" by R. P. Benedict, *Electro-Technology* (*New York*), October 1967, pp. 69-90, Vol. 80.

PROBLEMS

2.1 A skin diver descends to a depth of 60 ft in fresh water. What is the total pressure on his body?

2.2 If the specific weight of a liquid varies linearly with gage pressure, determine the pressure at a level of 1000 ft. Assume that the specific weight under atmospheric pressure is 40 lb/ft³ and that the variation of specific weight with pressure can be given as a change in specific weight of 10 lb/ft³ for every 100-psi change in pressure. Use a graphical method to solve this problem.

2.3 The pressure at a level of 100 ft in the Great Salt Lake is 46 psig. What is the specific weight of the lake water?

2.4 If the atmospheric pressure at some unknown elevation is half that at sea level, determine the elevation. Assume the temperature of the air to be constant and equal to 70°F.

2.5 Using equation (2.8) evaluate the height at which the pressure of the atmosphere will become zero. Would you expect this to be the case in the actual atmosphere?

2.6 An airplane uses a barometer as an altimeter. While approaching an airport the pilot is told that the local barometer stands at 29.8 in. Hg. Using this value to set his zero he reads a 1500 ft altitude. However, the instrument does not compensate for temperature and is correct only at 70°F. If the air is at 0°F, will he clear a 1400-ft mountain in his path?

2.7 A U-tube mercury manometer, open on one end, is connected to a pressure source. If the difference in the mercury levels in the tube is 6.5 in., determine the unknown pressure in psfa.

2.8 Two sources of pressure M and N are connected by a water-mercury differential gage as shown. What is the difference in pressure between M and N in psi?

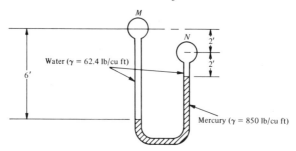

2.9 Determine the difference in pressure between A and B if the specific weight of water is 62.4 lb/ft³.

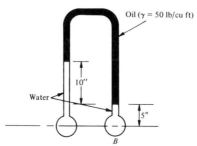

2.10 If the liquid in pipe B in problem 2.9 is carbon tetrachloride, whose specific weight is 99 lb/ft³, determine the pressure difference between A and B.

2.11 In a U-tube manometer one end is closed, trapping atmospheric air in the column. The other end is connected to a pressure supply of 5 psig. If the level of mercury in the closed end is 2 in. higher than that in the open end, what is the pressure of the trapped air?

2.12 A chimney depends on the difference in static pressure between the column of air on the inside and the column of air on the outside to provide the required draft. The specific weight of air at 70°F is 0.075 lb/ft³. Assume that the stack is 50 ft high and that the gas inside the stack is the same as air and is at 300°F. Determine the difference in pressure between the outside air column and the chimney column if the specific weight of each is constant.

2.13 To measure the draft in a chimney (see problem 2.12) an inclined manometer is connected so that the inclined leg (at 10 deg to the horizontal) is connected to the hot-gas side and the large reservoir is connected to the atmosphere. If the draft gage uses water, determine how far along the tube the water will travel. Use 0.1 psig as the differential being measured.

2.14 A manometer is connected to a pipe in which water is flowing as shown in Figure 2.11. Assume that the pipe is 20 in. in diameter and that the water specific weight is 50 lb/ft³ in the pipe. Due to heat transfer, the specific weight of the water in the manometer leg is 55 lb/ft³. If h_1 is 10 in. above the radius of the pipe, h is 5 in., and $\gamma = 850$ lb/ft³, determine the pressure at the center of the pipe.

2.15 A sensitive U-tube manometer is to be used to measure a small pressure difference. If the ends are each 2 in. i.d. and the tubing is $\frac{1}{4}$-in. i.d., what is the movement of the common liquid-surface in the tube for a pressure differential of 0.1 psi, if the fluids have specific weights of 50 and 60 lb/ft³, respectively?

2.16 For the arrangement shown, determine $p_A - p_B$.

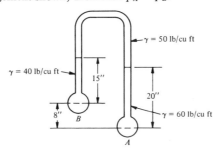

2.17 A cylindrical tank with its long axis horizontal is filled with a fluid to its center. Derive an expression for the force on the end of the cylinder in terms of R, the cylinder radius, and the specific weight of the fluid.

2.18 If a cylindrical tank is placed horizontally, determine the end force if it is 10 ft in diameter and the water level is at the center.

2.19 A trash rack, 5 ft wide, protects the intake to a power plant as shown in the accompanying sketch. In the event that the intake is blocked by ice and debris, what would be the total force acting on the rack and where would this force be applied?

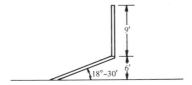

2.20 A vertical plate is used to dam a channel. If the plate has the shape shown, determine the total force on the plate. Assume that $\theta_1 = \theta_2$.

2.21 A vertical square plate is used as a dam in a channel. If the plate is 8 ft square and the liquid level on the downstream side is 4 ft high, determine the total force on the plate.

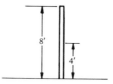

2.22 Determine the location of the resultant force in problem 2.20.

2.23 Determine the location of the resultant force in problem 2.21.

2.24 If the specific weight of air is 0.075 lb/ft^3, solve illustrative problem 2.13 accounting for the buoyant effect of the air.

2.25 A cylindrical log is 1 ft in diameter and 20 ft long. What weight of iron must be tied to one end of the log to keep it floating in an upright position in sea water with 18 ft of the log submerged? ($\gamma_{log} = 43.7$ lb/ft^3, $\gamma_{iron} = 493$ lb/ft^3, and $\gamma_{sea\ water} = 64.3$ lb/ft^3.)

2.26 A ship is placed in a dry dock and the dry dock is filled until the ship just floats. By measuring the volume of water in the dock with and without the ship in it, it is found that 10,000 ft^3 of sea water is displaced by the ship. If γ is 64.3 lb/ft^3, determine the weight of the ship.

2.27 A quarter circle is to be used as a gate, with the center of the quarter circle at the surface of the liquid. If the radius is 10 in. and the gate has a length of 10 in., determine the horizontal and vertical forces acting on the gate. Also determine the location of the point of application of these forces.

2.28 Determine the force F to keep the gate closed in the accompanying figure. Assume γ_W is the specific weight of the water and the gate is W ft wide.

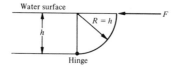

2.29 Determine the thickness of a cylinder subject to internal pressure if the pressure is 100 psia and the diameter is 4 ft. Assume an allowable stress of 20,000 psia in the material.

2.30 Solve problem 2.29 if the container is a sphere.

2.31 A pipe, 2 ft in diameter, has a wall $\frac{1}{4}$ in. thick. If the allowable stress of the material is 15,000 psi, what internal pressure can it sustain?

NOMENCLATURE

A = area

a = area

a, b, c = linear dimensions

b = barometric pressure

b = width

d = depth

F = force

h = height

$\bar{h}$ = height to center of gravity

I = moment of inertia

$\bar{I}$ = moment of inertia about the center of gravity

L = length

$\bar{L}$ = length to center of gravity

$\ln$ = logarithm to the base e

p = pressure

R = gas constant

R = radius

S = stress

T = absolute temperature

t = thickness

V = volume

v = specific volume

W = weight

x = linear dimension

y = height

Z = height

γ = specific weight

$\bar{\gamma}$ = average specific weight

Δ = small increment

θ = angle

μ = viscosity

ρ = density

$\sum$ = summation

Subscripts

A, B = locations

a = atmosphere

a = air

cp = center of pressure

g = gas

H = horizontal

u = unknown

V = vertical

W = water

x, y, z = coordinate directions

$1, 2$ = specific states

Chapter **3**

ENERGETICS OF
STEADY FLOW

3.1 Introduction

In Chapter 2 we studied fluids at rest, and by applying the requirement that static systems must be in equilibrium we were able to analyze any applications that occur in engineering. The next logical extension of the study of fluid mechanics is to those situations in which the fluid is flowing relative to its boundaries or to a fixed datum. A system has already been defined as a grouping of matter taken in any convenient or arbitrary manner. However, when dealing with fluids in motion it is more convenient to utilize the concept of an arbitrary volume in space, known as a control volume, that can be bounded by either a real or imaginary surface, known as the control volume. By correctly noting all of the forces acting on the fluid within the control volume, the energies crossing the control surface, and the mass crossing the control surface it is possible to derive mathematical expressions that will evaluate the flow of the fluid relative to the control volume. In this chapter we shall be concerned with fluids flowing steadily through the control volume, and for this type of system a monitoring station anywhere in the control volume will indicate no change in the fluid properties or energy quantities crossing the control surface with time. These quantities can and will vary from position to position in the control volume.

A word of caution is in order at this point—the concepts and equations that will be developed in this chapter are applicable to fluids flowing steadily. They should not be extended to unsteady (time-varying) cases unless the effects of time-varying terms, which are not considered in this chapter, are properly accounted for. Also, the flow will be assumed to one-dimensional; that is, the flow parameters are constant in planes perpendicular to the mean flow direction.

3.2 Conservation of Mass

As noted in Section 3.1, both energy and mass can enter and leave a control volume and cross the control surface of a system. Since we are considering steady flow systems, we can express the fact that the principle of conservation of mass for these systems requires that the mass of the fluid in the control volume at any time be constant. In turn this requires that the net mass flowing into the control volume must equal the net mass flowing out of the control volume at any instant of time. To express these concepts in terms of a given system, let us consider the system schematically in Figure 3.1.

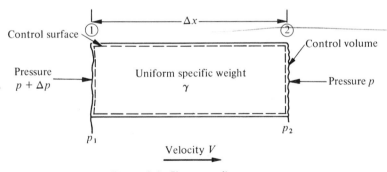

Figure 3.1. Elementary flow system.

Let us assume that at a certain time, fluid starts to enter the control volume by crossing the control surface at section 1 and that after a small interval of time the flowing fluid fills the pipe for a short distance Δx. If it is further assumed that in this short section of uniform pipe no heat is added or work is exchanged and that the specific weight stays constant, we can evaluate the amount of fluid that flowed in between sections 1 and 2. The weight contained between these sections is equal to the volume contained between the sections multiplied by the specific weight of the fluid. The volume is $A \, \Delta x$ and the specific weight is γ; therefore the contained weight is $\gamma A \, \Delta x$. The distance between stations, Δx, is simply $V \, \Delta t$, where V is the velocity of the fluid and Δt is the flow time required to fill the pipe between sections 1 and 2. Substituting this for Δ_x,

$$W = \gamma AV \, \Delta t \qquad (3.1a)$$

or

$$\dot{W} = \gamma AV \qquad (3.1b)$$

where $\dot{W}$ is the weight rate of flow per unit time, $W/\Delta t$. Also

$$W = \frac{AV}{v} \qquad (3.1c)$$

where the specific volume replaces the specific weight.

Equation (3.1) is known as the continuity equation, and as written for a pipe or duct we have made the assumption that the flow is normal to the pipe cross section and that the velocity V is either constant across the section or is the average value over the cross section of the pipe. The average value in this case is obtained by summing the products of velocity multiplied by the corresponding areas and dividing by the sum of all of the areas.

ILLUSTRATIVE PROBLEM 3.1 At the entrance to a steady flow device it is found that the pressure is 100 psia and that the specific weight of the fluid is 62.4 lb/ft³. If 10,000 ft³/min of fluid enters the system and the exit area is 2 ft², determine the mass flow rate and the exit velocity.

Solution. Refer to Figure 3.1. The weight flow rate is $\dot{W}$, which equals $\gamma AV = 62.4(10,000) = 624,000 \, \text{lb/min}$. Since this is the same at entrance and exit, the exit velocity is

$$\frac{AV}{A} = \frac{10,000}{2} = 5000 \, \text{ft/min}$$

3.3 Force, Mass, and Acceleration

Newton's second law of motion states that for a constant mass (m) the resultant force acting on a system is simply the product of its mass and its acceleration. Mathematically,

$$F = ma = \frac{W}{g}a \qquad (3.2a)$$

or

$$F = \frac{W}{g}\left(\frac{\Delta V}{\Delta t}\right)\lim \Delta t \rightarrow 0 \qquad (3.3)$$

since acceleration is the time rate of change of velocity. In equations (3.2a) and (3.3) the force is in pounds, the mass is in slugs, the weight is in pounds, and the acceleration is in feet per second per second. By transposing the terms in equation (3.3), we have for a constant mass system,

$$F(\Delta t) = \frac{W}{g}(\Delta V) \qquad (3.4)$$

where the left side is denoted as the impulse imparted to the body in the time Δt and the right side is the change of momentum of the body. Since both force and velocity are vector quantities, it should be noted that impulse and momentum are also vector quantities. In this chapter we shall be dealing with flow in one dimension and for simplicity will not use vector notation, but the student should note that vector quantities should be summed using the rules for vector addition.

Let us consider the one-dimensional system shown in Figure 3.2 and assume that it has a fixed volume, with fluid entering with velocity V_i and leaving with velocity V_o. In the time interval Δt, a quantity of fluid ΔW enters the control volume with velocity V_i. An equal amount of fluid must leave the control volume (since we have steady flow) with the assumed velocity V_o. The rate of momentum entering the system is

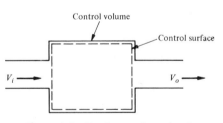

Figure 3.2. Simple one-dimensional steady flow system.

$$\text{Momentum in} = \frac{\dot{W}_i}{g}V_i \qquad (3.5a)$$

and the rate of momentum leaving the system is

$$\text{Momentum out} = \frac{\dot{W}_o}{g}V_o \qquad (3.5b)$$

Since $\dot{W}_i = \dot{W}_o = \dot{W}$ for steady flow, the net force in a given direction becomes,

$$F = \frac{\dot{W}}{g}(V_o - V_i) \tag{3.6}$$

Equation (3.6) is the momentum equation for the one-dimensional steady flow system.

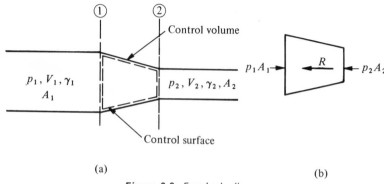

(a)   (b)

Figure 3.3. *Free body diagram*

As an example of the application of equation (3.6), consider the one-dimensional steady flow of a fluid in a circular pipe that has a uniform decrease in diameter, as shown in Figure 3.3. Due to the change in diameter there is a force exerted on the fluid by the walls of the channel. As shown in Figure 3.3, we denote the component of the wall force in the axial direction as R and also note that by symmetry there is no wall component perpendicular to the axial direction. Applying equation (3.6),

$$(p_1 A_1 - R - p_2 A_2) = \frac{\dot{W}}{g}(V_2 - V_1) \tag{3.7}$$

and

$$R = \left[p_1 A_1 + \frac{\dot{W}}{g}(V_1) \right] - \left[p_2 A_2 + \frac{\dot{W}}{g} V_2 \right] \tag{3.8}$$

The grouping $pA + [(\dot{W}/g)(V)]$ is also known as the thrust function. The internal force on the duct (R) is therefore the difference in the thrust function at the two sections.

ILLUSTRATIVE PROBLEM 3.2. A horizontal circular pipe is 1 ft in diameter and has a conical insert that decreases the diameter to $\frac{1}{2}$ ft. Determine the reaction on the nozzle when water flows at the rate of 4 ft³/sec in the pipe. Assume that the contraction causes the static pressure at the smaller diameter (section 2) to be 2.6 psi less than the static pressure at section 1, which is 20 psia.

Solution. Refer to Figure 3.3. From equation (3.8),

$$-R = \left[p_2 A_2 + \frac{\dot{W}}{g}(V_2)\right] - \left[p_1 A_1 + \frac{\dot{W}}{g}(V_1)\right]$$

where

$$V_1 = \frac{4}{\frac{\pi}{4}(1)^2} = 5.09 \text{ ft/sec}$$

$$V_2 = \frac{4}{\frac{\pi}{4}\left(\frac{1}{2}\right)^2} = 20.4 \text{ ft/sec}$$

$$\dot{W} = \gamma(AV) = 62.4(4) = 250 \text{ lb/sec}$$

$$A_1 = \frac{\pi}{4}(1)^2 = 0.785 \text{ ft}^2$$

$$A_2 = \frac{\pi}{4}\left(\frac{1}{2}\right)^2 = 0.197 \text{ ft}^2$$

$$p_2 = (p_1 - 2.6) \text{ psia} = 17.4 \text{ psia}$$

Therefore

$$+ R = -\left[(17.4)(144)(0.197) + \frac{250}{32.2}(20.4)\right]$$

$$+ \left[20 \times 144(0.7854) + \frac{250}{32.2}(5.09)\right]$$

$$= -494 - 158.5 + 2270 + 39.6 = +1657.1 \text{ lb}$$

Therefore $R = 1657.1$ lb acting on the pipe, as shown in Figure 3.3.

Since the difference in the thrust function is the internal force on the duct, it is the portion of the fluid reaction caused by the motion of the fluid. If other forces exist they must also be included if the net reaction on the pipe is desired. In illustrative problem 3.2 there is a force to the left due to the pressure of the atmosphere acting on the elements of the conical diffuser. This then equals the product of atmospheric pressure multiplied by $(A_1 - A_2)$.

Let us now consider the system in which mass can be added or removed. This case arises when jet planes or missiles are analyzed since their high rates of fuel consumption require that the varying mass of the system be considered. Restating equation (3.2a) and assuming m is a constant we have

$$F_{\text{res}} = ma = m\left(\frac{\Delta V}{\Delta t}\right) = \frac{\Delta(mV)}{\Delta t} \tag{3.2b}$$

where F_{res} is the resultant force acting on the body. Equation (3.2b) is based upon our assumption of constant mass, and it is now necessary to consider the case of a body of variable mass. Let the mass of the body at any time t be m, the velocity of the body be V, the initial velocity of the body being picked up at time t be V_i, and F_{ext} be the external force on the body at time t exclusive of any force due to jet action.

During an infinitesimal length of time from t to $t + \Delta t$, the mass picked up is $(\Delta m/\Delta t)\,\Delta t$, and its increase in momentum is $(V - v)(\Delta m/\Delta t)\,\Delta t$. During time Δt the increase in momentum of mass m is $m\,\Delta V$. Since impulse equals change in momentum,

$$F_{ext}\,\Delta t = m\,\Delta V + (V - v)\left(\frac{\Delta m}{\Delta t}\right)\Delta t \tag{3.9}$$

Dividing by Δt,

$$F_{ext} = m\left(\frac{\Delta V}{\Delta t}\right) + (V - v)\frac{\Delta m}{\Delta t} \tag{3.10}$$

Note that $m(\Delta V/\Delta t)$ is the rate of change of momentum of the instantaneous mass, while $(V - v)\,\Delta m/\Delta t$ is the rate of change of momentum of the added mass. Hence,

$$F_{ext} = \text{Rate of change of momentum of instantaneous mass} +$$
$$\text{Rate of change of momentum of added mass} \tag{3.11}$$

To compare (3.2b) with (3.11) it is necessary to expand the term $\Delta(mV)/\Delta t$. Performing this operation we have

$$\frac{\Delta(mV)}{\Delta t} = \frac{(m + \Delta m)(V + \Delta V)}{\Delta t} = m\left(\frac{\Delta V}{\Delta t}\right) + V\left(\frac{\Delta m}{\Delta t}\right) \tag{3.12}$$

Substituting (3.12) into (3.10),

$$F_{ext} = \Delta\left(\frac{mV}{\Delta t}\right) - V_i\left(\frac{\Delta m}{\Delta t}\right) \tag{3.13}$$

Therefore equation (3.2b) is valid only when the mass picked up is initially at rest and $F_{res} = F_{ext}$.

The mass being picked up by the body at time t exerts on the body a jet force $F_{jet} = (V - v)\,\Delta m/\Delta t$ that acts in a direction opposite to V. Inserting this into equation (3.10) yields

$$F_{\text{ext}} = m\left(\frac{\Delta V}{\Delta t}\right) + F_{\text{jet}} \qquad (3.14)$$

Equation (3.14) can also be written as,

$$F_{\text{res}} = m\left(\frac{\Delta V}{\Delta t}\right) \qquad (3.15)$$

with $F_{\text{res}} = F_{\text{ext}} - F_{\text{jet}}$. Hence, equation (3.2b) is also valid for a body of variable mass, provided $F_{\text{res}} = F_{\text{ext}} - F_{\text{jet}}$. By a similar analysis it can be shown that these equations also hold when the body is discarding mass.

From the discussion above we can conclude that $F = ma$ can be applied to those situations where the mass is either constant or variable provided that the resultant force includes any jet force on the body and $F = \Delta(mv)/\Delta t$ is valid for variable mass only when the added mass is initially at rest or the discarded mass is finally at rest.

It should be noted, however, that the resultant force is equal to the rate of change of the momentum for both constant and variable mass provided that the resultant force does not include any jet force and that the rate of change of momentum is calculated separately for the instantaneous mass and the added [or discarded] mass.

ILLUSTRATIVE PROBLEM 3.3. In a turbojet engine, air enters and is compressed, fuel is added and burned, energy is removed by a turbine, and finally the combustion gases are exhausted. For such an engine let V_1 be the inlet air velocity relative to the engine, $\dot{W}_A$ the weight rate of air flow, $\dot{W}_f$ the weight rate of fuel flow, and V_2 the exhaust velocity relative to the engine. Assume that the inlet pressure, atmospheric pressure, and the exhaust pressure are all equal and that the inlet and exit areas are equal and determine the thrust of the engine.

Solution. Since V_1 and V_2 are specified relative to the engine, we may con-

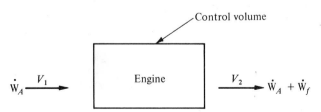

sider the engine fixed and the inlet and exhaust stream velocities as absolute velocities. The momentum in is $\dot{W}_A/g(V_1)$ and the momentum out is $(\dot{W}_A +$

$\dot{W}_f/g(V_2)$. Since the pressures and areas have been assumed equal we can directly write the thrust as

$$F = \frac{\dot{W}_A + \dot{W}_f}{g}V_2 - \frac{\dot{W}_A}{g}V_1$$

since it is simply the rate of change of momentum into and out of the control volume.

3.4 Energy, Work, and Heat

In this text we shall define the work done by a force as the product of the displacement of the body multiplied by the component of the force in the direction of the displacement. Thus in Figure 3.4a the displacement of the body on

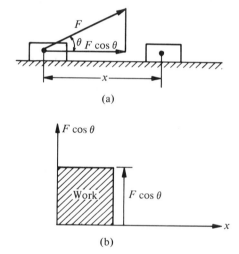

(a)

(b)

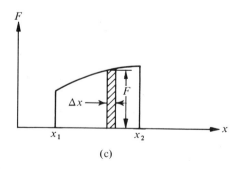

(c)

Figure 3.4. Work.

the horizontal plane is x and the component of the force in the direction of the displacement is $F \cos \theta$. The work done is therefore $(F \cos \theta)(x)$. The constant force $(F \cos \theta)$ is plotted as a function of x in Figure 3.4b, and it will be noted that the resulting figure is a rectangle. The area of this rectangle (shaded) is equal to the work done, since it is $(F \cos \theta)(x)$. If the force varies so that it is a function of the displacement, it is necessary to consider the variation of force with displacement in order to find the work done. Figure 3.4c shows a general plot of force as a function of displacement. If the displacement is subdivided into many small parts, Δx, and for each of these small parts F is assumed to be very nearly constant, it is apparent that the sum of the small areas (F) (Δx) will represent the total work done when the body is displaced from x_1 to x_2. Thus the area under the curve of F as a function of x represents the total work done if F is the force component in the direction of x.

At this point let us define the term *energy* in terms of work. Energy can be described (in an elementary manner) as the capacity to do work. At first it may appear that this definition of energy is too restrictive when applied to electrical and magnetic systems. Yet in all instances the observed effects on a system can (ideally) be converted to mechanical work. For example, the motion of a charged particle in a field involves energy changes since work must have been done on the particle either to create the motion of the particle or the change its motion. Since work has been defined as the product of a force multiplied by a displacement, it is not stored in a system. It represents energy that must be crossing the boundaries (real or imaginary) of the system and can properly be denoted to be energy in transition.

To distinguish between the transfer of energy as work to or from a system, we shall adopt the convention that work done by a system on its surroundings is positive and work done by the surroundings on the system represents negative work. For the student it is best to think of useful work out of a system as being a desirable quantity and therefore as positive. For example, consider a spring that is compressed by a force. If the spring is the system, the work is negative by convention. If the external force system acting on the spring is the system, the work is positive by convention.

The work that a system can perform on its surroundings is not an intrinsic property of the system. Work is a function of the path taken and is a transitory effect that is not stored in the system and that is not a property of the system. There is one process, however, that does permit the evaluation of the work done since the path is uniquely defined. This process is called the quasi-static process, and we shall find it useful in subsequent work. The quasi-static process is a process carried out infinitely slowly so that it is in equilibrium at all times. Before considering several processes, let us first discuss another form of energy that can cross the boundaries of a system, namely, *heat*.

When a *heat* interaction occurs in a system, it is observed that two distinct events occur. The first is that an interchange of energy has occurred between the system and its surroundings. The second is that this would not have taken place if there were no temperature difference between the system and its surroundings. We may therefore define *heat* as the energy in transition crossing the boundaries of a system due to a temperature difference between the system and its surroundings. In this definition of heat the transfer of mass across the boundaries of the system is excluded. It should be noted that this definition of heat indicates a similarity between heat and work. Both are energies in transition and both are not properties of the system in question. Just as was the case for work, we can have quasi-static heat transfer to or from a system. In this case the difference in temperature between the system and its surroundings can be only an infinitesimal amount at any time. Once again it is necessary to adopt a convention for the energy interchanged by a system with its surroundings as heat. We shall use the convention that heat to the system from its surroundings is positive and that heat out of the system is negative. To learn these conventions it is convenient to think of the conventional situation where heat is transferred to a system in order to obtain useful work from the system. This sets the convention that heat into a system is positive and work out of the system is also positive. *Positive* in this sense means either desirable or conventional from the viewpoint of conventional power cycles.

Since work and heat are both forms of energy in transition, it follows that the units of work should be capable of being expressed as heat units and vice versa.

At present this conversion factor is defined as 778.16 ft lb/Btu and is conventionally given the symbol J. In this text we shall use the symbol J to be 778 ft lb/Btu since this is sufficiently accurate.

3.4a Potential Energy

Let us consider the following problem: A body of mass m is in a locality where the local gravitational field is constant and equal to g. A force is applied to the body, and it is raised a distance Z from its initial position. The force will be assumed to be only infinitesimally greater than the mass so that the process is carried out as a quasi-static process. In the absence of electrical, magnetic, and other extraneous effects, determine the work done on the body. The solution to this problem is obtained by noting that the equilibrium of the body requires that a force must be applied to it equal to its weight. The weight of the body is simply W. In moving through a distance Z the work done by this force will therefore equal

$$\text{Work} = WZ \qquad (3.16)$$

The work done on the body can be returned to the external environment by simply reversing this quasi-static process. We therefore conclude that this system has had work done on it equal to WZ and that, in turn, the system has stored in it an amount of energy in excess of the amount it had in its initial position. The energy added to the system in this case is called potential energy. Thus

$$\text{Potential energy (P.E.)} = WZ \qquad (3.17)$$

A feature of importance of potential energy is that a system can be said to possess potential energy only with respect to an arbitrary initial or datum plane.

3.4b Kinetic Energy

Let us consider another situation in which a body of mass m is at rest on a frictionless plane. If a force F is applied to the mass, it will be accelerated in the direction of the force. After moving through a distance Δx, the velocity of the body will have increased from V_1 to V_2. The only effect of the work done on the body will be an increase in its velocity. See Figure 3.5

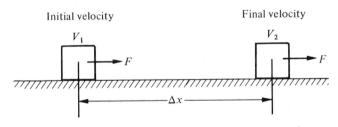

Figure 3.5

To solve this situation we shall apply Newton's second law as the impulse-momentum relation. Thus

$$F(\Delta t) = \frac{W}{g}(\Delta V) \qquad (3.18)$$

where Δt is the time interval during which the force is applied and ΔV the change in velocity during this time interval. The distance traveled during this time Δx is the average velocity during the time interval multiplied by the time interval. Mathematically,

$$\Delta x = \bar{V}(\Delta t) \tag{3.19}$$

where $\bar{V}$ is the average velocity. If the value of Δt from equation (3.19) is substituted into equation (3.18),

$$F\left(\frac{\Delta x}{\bar{V}}\right) = \frac{W}{g}(\Delta V) \tag{3.20a}$$

or

$$F(\Delta x) = \frac{W}{g}\bar{V}(\Delta V) \tag{3.20b}$$

The change in velocity V is equal to $V_2 - V_1$, and the average velocity $\bar{V}$ is equal to $\left(\frac{V_1 + V_2}{2}\right)$. These relations can be placed into (3.20b) to obtain

$$F(\Delta x) = \frac{W}{g}\left(\frac{V_1 + V_2}{2}\right)(V_2 - V_1) \tag{3.21}$$

and

$$F(\Delta x) = \frac{WV_2^2}{2g} - \frac{WV_1^2}{2g} \tag{3.22}$$

It will be noted that the left side of equation (3.22) is the work done on the system by the force F. The terms on the right side of equation (3.22) are called the kinetic energy (K.E.) of the body at the beginning and end of the process. Thus equation (3.22) expresses the energy possessed by a body of mass W/g having a velocity V relative to a stationary reference. It is usual to consider the earth as being the reference.

3.4c Internal Energy

To this point we have considered the energy in a system that arises from the work done on the system. However, a body possesses energy by virtue of the molecular motion of the molecules of the body. In addition to the energy possessed by the body due to the motion of its molecules, it also possesses energy due to the internal attractive and repulsive forces between molecules. These forces give rise to the storage of energy internally as potential energy. The energy that a body possesses from all such sources is called the internal energy of the body and is designated by the symbol U. Per unit mass W/g the specific internal energy is denoted by the symbol u, where $Wu/g = U$. From a practical standpoint, the measurement of the absolute internal energy of a system in a given state presents an insurmountable problem and is not essential to our study. We shall be concerned with changes in internal energy, and the arbitrary datum for the zero of internal energy will not enter these problems. Internal energy is a property of a system and is not dependent on the path taken to place the system in a given configuration.

Just as it is possible to distinguish the various forms of energy, such as work and heat in a mechanical system, it is equally possible to distinguish the various forms of energy associated with electrical, chemical, and other systems. For the purposes of this text, these forms of energy other than work and heat will not be considered. The student is cautioned that should a system include any forms of energy other than mechanical, these items must be included. Thus the energy that is dissipated in a resistor as heat when a current flows through it must be taken into account when all of the energies of an electrical system are being considered.

3.5 Steady Flow Energy Equation

The First Law of Thermodynamics is the statement of the conservation of energy; it can be expressed by stating that energy can be neither created nor destroyed but only converted from one form to another. Since we shall not be concerned with nuclear reactions, it will not be necessary to invoke the interconvertibility of energy and mass in this study.

When a fluid is caused to flow in a system it is necessary that somewhere in the system work must have been supplied. This work can be evaluated by referring to Figure 3.1. The work is the product of the unbalanced force multiplied by the displacement. The unbalanced force is $(p_1 - p_2)A$, and the displacement is Δx; therefore,

$$\text{Total work} = (p_1 - p_2)A\,\Delta x$$

For the time that it takes the fluid to traverse the distance Δx, $[\gamma_1(AV_1)\,\Delta t]$ lb of fluid must have been displaced. The work per pound of fluid is then,

$$W = \frac{(p_1 - p_2)(A\,\Delta x)}{\gamma_1 AV_1(\Delta t)} \tag{3.23}$$

Since the weight flow rate must be a constant (otherwise fluid would be added to or depleted from the system),

$$\gamma_1 A_1 V_1 = \gamma_2 A_2 V_2 \quad \text{or} \quad \gamma V_1 = \gamma V_2 \quad \text{if } A_1 = A_2 \qquad (3.24)$$

Dividing each term in the numerator of (3.23) by the denominator and utilizing equation (3.24),

$$\frac{W}{\gamma A V} = \frac{p_1 \, \Delta x}{\gamma_1 V_1 \, \Delta t} - \frac{p_2 \, \Delta x}{\gamma_2 V_2 \, \Delta t} \qquad (3.25)$$

But Δx is closely either $V_1 \, \Delta t$ or $V_2 \, \Delta t$. Using this and the fact that the specific volume v is the reciprocal of the specific weight γ yields

$$\text{Flow work (ft lb/lb)} = p_1 v_1 - p_2 v_2 = \frac{p_1}{\gamma_1} - \frac{p_2}{\gamma_2} \qquad (3.26)$$

The interpretation of equation (3.26) is that the difference in the p/γ terms represents the amount of work that is done on a system to introduce a unit mass into it minus the work done by the mass on its environment as it leaves the system.

In the foregoing derivations, certain assumptions were made, and these are repeated here for emphasis. The term *steady*, when applied to a flow situation, means that the condition at any section of the system is independent of time. Even though the velocity, specific weight, and temperature of the fluid can vary in an arbitrary manner across the stream, they are not permitted to vary with time. The weight entering the system per unit time must equal the weight leaving the system in the same period of time; otherwise the system would either store or be depleted of fluid.

To summarize, we have identified six energy terms that apply to various situations:

Item	Value (ft lb/lb)
Potential energy	Z
Kinetic energy	$V^2/2g$
Internal energy	Ju
Flow work	p/γ or pv
Work	W
Heat	Jq

Figure 3.6 shows such a system, where it is assumed that each form of energy can both enter and leave the system. At entrance $\dot{W}$ lb of fluid/sec enter, and

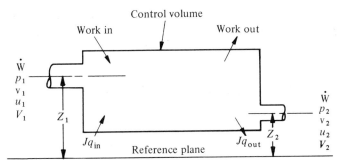

Figure 3.6. *Steady flow system.*

the same amount leaves at the exit. At the entrance the fluid has a pressure of p_1, a specific volume of v_1, an internal energy of u_1, and a velocity of V_1. At the exit we have similar quantities expressed as p_2, v_2, u_2, and V_2. The fluid enters and leaves at different elevations, and both work and heat cross the boundary in both directions. We express the conservation of energy for this steady flow system by stating that all of the energy into the system must equal all of the energy leaving the system. Setting up this energy balance:

	Energy In (ft lb/lb)	**Energy Out (ft lb/lb)**
Potential energy	Z_1	Z_2
Kinetic energy	$V_1^2/2g$	$V_2^2/2g$
Internal energy	Ju_1	Ju_2
Flow work	p_1v_1 (or p_1/γ_1)	p_2v_2 (or p_2/γ_2)
Work	W_{in}	W_{out}
Heat	Jq_{in}	Jq_{out}

Equating all of these terms;

$$Z_1 + \frac{V_1^2}{2g} + Ju_1 + \frac{p_1}{\gamma_1} + W_{\text{in}} + Jq_{\text{in}}$$

$$= Z_2 + \frac{V_2^2}{2g} + Ju_2 + \frac{p_2}{\gamma_2} + W_{\text{out}} + Jq_{\text{out}} \tag{3.27}$$

Equation (3.27) is quite general and is a consequence of the principle of conservation of energy for this system.

If at this point we note that the heat and work terms can each be combined to form individual terms of net heat and net work and if we take proper cognizance of the mathematical signs of these net terms, we can rewrite equation (3.27) as

$$Z_1 + \frac{V_1^2}{2g} + Ju_1 + \frac{p_1}{\gamma_1} + Jq = Z_2 + \frac{V_2^2}{2g} + Ju_2 + \frac{p_2}{\gamma_2} + W \tag{3.28}$$

In equation (3.28) all terms are written as foot pounds per pound and q and W represent net values. If we now further note that the grouping of terms $u + pv$ appears on both sides of equation (3.28) (since p/γ equals pv) and that this combined term is a property, we obtain

$$h = u + \frac{pv}{J} \tag{3.29}$$

where h is given the name *enthalpy*. Regrouping terms in equation (3.28) and using equation (3.29),

$$q - \frac{W}{J} = (h_2 - h_1) + \left(\frac{Z_2 - Z_1}{J}\right) + \left(\frac{V_2^2 - V_1^2}{2gJ}\right) \qquad \text{in Btu/lb} \tag{3.30}$$

The student should notice that equations (3.27) through (3.30) are essentially the same. To intelligently apply these equations it is important that each of the terms be completely understood. Although these equations are not difficult, the student may encounter difficulty at this point. Most of this difficulty arises from a lack of understanding of these energy equations and the basis for each of the terms in them.

Before illustrating the use of these energy equations, let us consider the following situation. A fluid is flowing steadily in a device in which it undergoes a compression. Let us further assume that this process is quasi-static. For this process the work term can be written as the sum of the work done on the fluid plus the flow work. Kinetic and potential energy terms will be assumed to be negligible. Thus the work for this process becomes

$$p_1 v_1 - p_2 v_2 - p \, \Delta v \tag{3.31}$$

Let us now plot a pv diagram for this fluid, as shown in Figure 3.7. As shown in Figure 3.7a, the term $p \, \Delta v$ is the area under the curve between the limits 1 and 2. By subtracting $p_1 v_1$ and adding $p_2 v_2$ (graphically), the result is $v \, \Delta p$ and is portrayed in Figure 3.7b. Mathematically,

$$- v \, \Delta p = p_1 v_1 - p_2 v_2 - p \, \Delta v \tag{3.32}$$

For the flow system, flow work of an amount $p_1 v_1$ enters the system, and flow work of an amount $p_2 v_2$ leaves the system. In between, work is done on the working fluid. We can therefore interpret the meaning of the areas on Figure 3.7 or the terms of equation (3.32) as follows: The work of the quasi-static frictionless flow system (neglecting kinetic and potential energy terms) is the

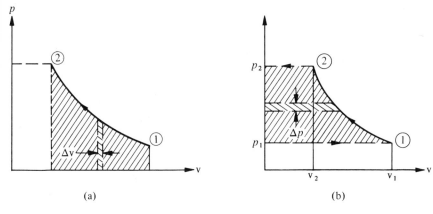

(a)   (b)

Figure 3.7

algebraic sum of the work to induct the fluid into the the system plus the work of compression (or expansion) less the flow work to deliver the fluid to the downstream exit. It is sometimes convenient to think of $p\,\Delta v$ as the work of *compression* (or expansion) and $v\,\Delta p$ as the work of *compressor* (or expander). It is extremely important to note that both $p\,\Delta v$ and $v\,\Delta p$ evaluate work only for a quasi-static process or system. For systems that are real they do not evaluate work. However, for a large class of systems they approximate the work terms, and their use is justified both from this approximation and the relative ease with which they can be evaluated.

3.6 The Bernoulli Equation

In fluid mechanics, use is frequently made of an equation known as the Bernoulli equation. In Chapter 4 this equation will be studied in detail; however, it is in order to undertake a brief discussion of it at this point. Let us consider a system in which the flow is steady, there is no change in internal energy, no work is done on or by the system, no energy as heat crosses the boundaries of the system, and the fluid is incompressible. We shall further assume that all processes are ideal in the sense that they are frictionless. For this system the energy equation reduces to the following in mechanical units of foot pounds per pound:

$$Z_1 + \frac{V_1^2}{2g} + p_1 v_1 = Z_2 + \frac{V_2^2}{2g} + p_2 v_2 \qquad (3.33a)$$

or

$$Z_1 + \frac{V_1^2}{2g} + \frac{p_1}{\gamma_1} = Z_2 + \frac{V_2^2}{2g} + \frac{p_2}{\gamma_2} \qquad (3.33b)$$

where $\gamma_1 = \gamma_2$. Equation (3.33b) is the usual form of the Bernoulli equation. Each term can represent a height since the dimension of foot pounds per pounds corresponds numerically (and dimensionally) to a height. Hence, the term *head* is frequently used to denote each of the terms in equation (3.33b). From the derivation of this equation we note that its use is restricted to those situations where the flow is steady, there is no friction, no shaft work is done on or by the fluid, the flow is incompressible, there is no change in internal energy during the process, and there is no heat transfer to or from the system.

3.7 Specific Heat and Some Applications of the Energy Equation

The term *specific heat* will be defined as the ratio of the energy *as heat* transferred during a particular process per unit mass of fluid involved divided by the corresponding change of temperature of the fluid that occurs during this process. Since heat can be transferred to or from a fluid and since algebraic signs have been adopted for the direction of heat transfer, it is entirely possible for a process to have a negative specific heat. The student should not confuse the specific heat of a process with the specific heat property.

For two processes the definition of specific heat given above is of importance because these processes serve to define a new property of a fluid. These processes are the constant pressure process and the constant volume process. It will be found from the earlier work of this chapter that the constant pressure process can be characterized by the energy equation $q = h_2 - h_1 = \Delta h$. This is found for the steady flow process without change in elevation, in the absence of external work, and with negligible changes in kinetic energy. For a nonflow constant pressure process that corresponds to closed systems one obtains the same result. Thus for any constant pressure process the specific heat is defined as

$$C_p = \left(\frac{q}{\Delta T}\right)_p = \left(\frac{\Delta h}{\Delta T}\right)_p \qquad (3.34)$$

where the subscript p indicates a constant pressure process.

For the constant volume process (which can be only a nonflow process in a closed system) it can be shown that $q = u_2 - u_1 = \Delta u$. Therefore,

$$C_v = \left(\frac{q}{\Delta T}\right)_v = \left(\frac{\Delta u}{\Delta T}\right)_v \qquad (3.35)$$

where the subscript v indicates constant volume. These specific heats are properties of the fluid and depend only on the state of the fluid.

ILLUSTRATIVE PROBLEM 3.4. A gas flows steadily in a pipe so that at one section of pipe its pressure and temperature are 100 psia and 950°F. At a second section of the pipe the pressure is 76 psia and the temperature is 580°F. The specific volume of the inlet gas is 4.0 ft³/lb, and at the second section it is 3.86 ft³/lb. Assume that the specific heat at constant volume is 0.32 Btu/lb°F. If no shaft work is done and if the velocities are small, determine the magnitude and direction of the heat transfer.

Solution. Drawing a sketch and writing an energy equation,

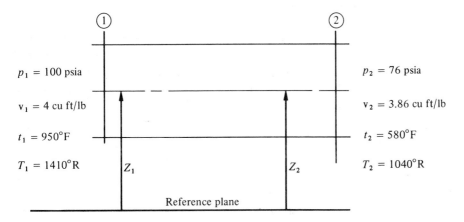

$$Z_1 + \frac{V_1^2}{2g} + Ju_1 + \frac{p_1}{\gamma_1} + Jq = Z_2 + \frac{V_2^2}{2g} + Ju_2 + \frac{p_2}{\gamma_2} + W$$

Based upon the statement of the problem,

$$(Z_1 - Z_2) \to 0$$
$$\left(\frac{V_1^2 - V_2^2}{2g}\right) \to 0$$
$$W \to 0$$

which reduces the energy equation to

$$u_1 + \frac{p_1}{\gamma_1} + Jq = u_2 + \frac{p_2}{\gamma_2}$$

But equation (3.35) permits us to express the internal energy change in terms of the temperature change as

$$u_2 - u_1 = C_v(T_2 - T_1) \qquad \text{for constant } C_v$$

Therefore,

$$q = C_v(T_2 - T_1) + \frac{p_2 v_2}{J} - \frac{p_1 v_1}{J} \qquad \text{in units of Btu/lb}$$

Inserting numerical quantities,

$$q = 0.32(1040 - 1410) + \frac{76(144)(3.86)}{778} - \frac{100(144)(4)}{778}$$

$$= -138.2 \text{ Btu/lb}$$

Thus 138.2 Btu/lb are transferred *from* the fluid.

ILLUSTRATIVE PROBLEM 3.5. A turbine receives air at 150 psia and 1000°R and discharges to a pressure of 15 psia. The final temperature at discharge is 600°R. If C_p of air can be taken to be constant over this temperature range and equal to 0.24 Btu/lb°F, determine the work output of the turbine per pound of working fluid. At inlet conditions the specific volume is 2.47 ft³/lb, and at outlet conditions it is 14.8 ft³/lb.

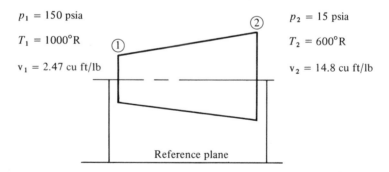

p_1 = 150 psia

T_1 = 1000°R

v_1 = 2.47 cu ft/lb

p_2 = 15 psia

T_2 = 600°R

v_2 = 14.8 cu ft/lb

Reference plane

Solution. Once again a sketch of the problem is useful. Writing the energy equation in terms of thermal units,

$$\frac{Z_1}{J} + \frac{V_1^2}{2gJ} + u_1 + \frac{p_1 v_1}{J} + q = \frac{Z_2}{J} + \frac{V_2^2}{2gJ} + u_2 + \frac{p_2 v_2}{J} + \frac{W}{J}$$

and noting that for this machine the difference in elevation between inlet and outlet would be reasonably small, that large differences in kinetic energy at the inlet and outlet would represent poor design, and that this device would normally be insulated to prevent heat losses to the surroundings, we can simplify matters considerably. Thus we obtain

$$u_1 + \frac{p_1 v_1}{J} = u_2 + \frac{p_2 v_2}{J} + \frac{W}{J}$$

or

$$h_1 = h_2 + \frac{W}{J}$$

and

$$h_1 - h_2 = \frac{W}{J}$$

From equation (3.34)

$$h_1 - h_2 = C_p(T_1 - T_2)$$

Thus

$$0.24(1000 - 600) = \frac{W}{J} = 96 \text{ Btu/lb.}$$

Illustrative problems 3.4 and 3.5 serve to illustrate two applications of the first law applied to steady flow systems as well as some considerations regarding specific heats. To further illustrate the utility of the energy equation when applied to steady flow systems, we shall consider the following two problems.

ILLUSTRATIVE PROBLEM 3.6. Steam is expanded in a nozzle from an initial enthalpy of 1220 Btu/lb to a final enthalpy of 1100 Btu/lb. If the initial velocity of the steam is negligible, what is the final velocity? If the initial velocity is 1000 ft/sec, what is the final velocity?

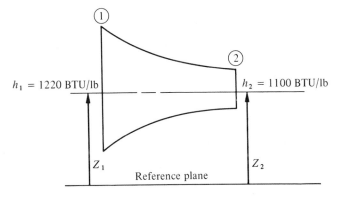

Solution. A nozzle is essentially a short device in which no work is added while the fluid expands. The energy equation is

$$Z_1 + \frac{V_1^2}{2g} + u_1 + \frac{p_1}{\gamma_1} + Jq = Z_2 + \frac{V_2^2}{2g} + u_2 + \frac{p_2}{\gamma_2} + W$$

For this device differences in elevation are negligible; no work is done on or by the fluid; friction will be taken to be negligible and due to the speed of the fluid flowing, and the short length of the nozzle heat transfer to or from the surroundings will also be taken to be negligible. Under these circumstances,

$$Ju_1 + \frac{p_1}{\gamma_1} + \frac{V_1^2}{2g} = Ju_2 + \frac{p_2}{\gamma_2} + \frac{V_2^2}{2g}$$

or

$$J(h_1 - h_2) = \frac{V_2^2}{2g} - \frac{V_1^2}{2g}$$

1. For negligible entering velocity,

$$V_2 = \sqrt{2gJ(h_1 - h_2)}$$

Substituting the data of the problem;

$$V_2 = \sqrt{2 \times 32.2 \times 778(1220 - 1100)} = 2450 \text{ ft/sec}$$

2. If the initial velocity is appreciable,

$$J(h_1 - h_2) + \frac{V_1^2}{2g} = \frac{V_2^2}{2g}$$

Again inserting numerical values,

$$1220 - 1100 + \frac{(1000)^2}{2 \times 32.2 \times 778} = \frac{V_2^2}{2 \times 32.2 \times 778}$$

$$120 + 20 = \frac{V_2^2}{2 \times 32.2 \times 778}$$

and

$$V_2 = 2640 \text{ ft/sec}$$

Note that in this part of the problem the entering velocity was nearly 40 percent of the final velocity, yet neglecting the entering velocity makes only a 10 percent error in the answer. It is quite common to neglect the entering velocity in many of these problems.

ILLUSTRATIVE PROBLEM 3.7. Air is flowing steadily in a pipe. At some section of the pipe there is an obstruction that causes an appreciable local pressure loss. Derive the energy equation for this process after flow has become uniform in the downstream section of the pipe. This process is known as a throttling process and is characteristic of valves and orifices placed in pipe lines.

Solution. For the conditions of this problem we can take differences in elevation to be negligible, differences in kinetic energy terms will be relatively small, and no work or heat crosses the system boundary. The complete energy equation is

$$Z_1 + \frac{V_1^2}{2g} + Ju_1 + \frac{p_1}{\gamma_1} + Jq = Z_2 + \frac{V_2^2}{2g} + Ju_2 + \frac{p_2}{\gamma_2} + W$$

and with the assumptions made for this process,

$$u_1 + \frac{p_1}{\gamma_1 J} = u_2 + \frac{p_2}{\gamma_2 J} \qquad \text{(in appropriate units)}$$

or

$$h_1 = h_2$$

In words, a throttling process is carried out at constant enthalpy. One assumption should be verified; that is, the kinetic energy differences at inlet and outlet are indeed negligible. This can easily be done by using the constant enthalpy condition and using the final properties of the fluid and the cross-sectional area of the duct to determine an approximate final velocity.

3.8 The Second Law of Thermodynamics

Thus far we have considered various forms of energy (as well as energy in transition, as both work and heat) without regard to any limitations on these quantities. It has been assumed that work and heat are mutually interchangeable forms of energy, and it may have appeared to the student that the distinction made between these quantities was arbitrary and possibly not necessary. The Second Law of Thermodynamics expresses these limitations. Since the interest in this text is the application of physical principles to the mechanics of fluids, we shall state the Second Law and also state the developments from it that will be required later on in this text. The interested student is referred to any standard text on thermodynamics for a complete discussion of the Second Law. The Second Law of Thermodynamics is an expression of the empirical fact that all forms of energy are not necessarily equivalent in their ability to perform useful work. For the present the Clausius expression that "Heat cannot, of itself, pass from a lower to a higher temperature" will serve to fully express the Second Law. Using this axiom (or statement of the Second Law) and applying it to processes and engine cycles, it is possible to arrive at a new property called *entropy*. Without exploring the derivation, meaning, and applications of this new property, we shall simply accept it as a working tool in the same manner that pressure or temperature are accepted. For a quasi-static process carried out in the absence of friction, temperature differences, combination, or internal molecular inelastic action (this process is called a reversible process), the change in entropy is defined as,

$$\Delta s = \left(\frac{q}{T}\right)_{\text{reversible process}} \tag{3.36}$$

The energy equation for a reversible *nonflow* process can be written as

$$q = \Delta u + \frac{p\,\Delta v}{J} \tag{3.37}$$

Since we are considering a reversible process, Δq can be evaluated from equation (3.36):

$$T\,\Delta s = \Delta u + \frac{p\,\Delta v}{J} \tag{3.38}$$

Notice that each term of equation (3.38) represents a property of the fluid. By now referring to the definition of enthalpy,

$$\Delta h = \Delta u + \frac{\Delta(pv)}{J} \tag{3.39}$$

and using equation (3.38)

$$T\,\Delta s = \Delta h - \frac{\Delta(pv)}{J} \tag{3.40}$$

If the $\Delta(pv)$ term is expanded and higher-order products are neglected, it can be shown that $\Delta(pv) = (\Delta p)v + p(\Delta v)$. Therefore,

$$T\,\Delta s = \Delta h - \frac{v\,\Delta p}{J} \qquad (3.41)$$

Equations (3.36) to (3.41) will be found to be useful in the study of compressible flow in Chapter 7.

Because of their uniqueness, several thermodynamic processes have been given special names. Thus a process carried out without the transfer of heat is known as an adiabatic process. A process carried out at constant temperature is called an isothermal process. Similarly a reversible adiabatic process is called an isentropic process since it is carried out with zero change in entropy. These three processes will be studied in conjunction with the flow of a compressible fluid in Chapter 7.

3.9 Ideal Gas

An equation relating the variables of pressure, temperature, and volume of a fluid is called the equation of state for the fluid. The simplest equation of state for a gas, which is also reasonably accurate for most engineering applications, is

$$pv = RT \qquad (3.42)$$

where p is the pressure in psfa, v the specific volume in cubic feet per pound, R a constant depending on the gas, and T the absolute temperature in degrees Rankine. In these units R is closely 1545 divided by the molecular weight. Therefore, R for $N_2 = 1545/28 = 55.2$, R for H_2O is $1545/18 = 85.8$, and R for air is closely $1545/29 = 53.3$. Equation (3.42) relates the properties of the fluid to each other at any point regardless of how the particular point is arrived at. Thermodynamically, it is necessary to know the path taken if we are to be able to express the energy equation and Second Law explicity for a process. In general this is quite difficult, but for the ideal gas certain processes have paths that can readily be determined. The path of the isentropic process can be shown to follow

$$pv^k = \text{Constant} \qquad (3.43)$$

where k is the ratio of C_p/C_v. For the general process (sometimes called the polytropic process) the path equation is sometimes written as

$$pv^n = \text{Constant}$$

where n is a characteristic of the process. For example, for a constant volume process $n = 1$, etc.

From the definition of enthalpy we have

$$\Delta h = \Delta u + \frac{\Delta(pv)}{J} \tag{3.44}$$

Applying equations (3.34) and (3.35) and (3.42),

$$C_p(\Delta T) = C_v \, \Delta T + \frac{R \, \Delta T}{J} \tag{3.45a}$$

or

$$C_p - C_v = \frac{R}{J} \tag{3.45b}$$

Using the identity that $C_p/C_v = k$,

$$C_v = \frac{R}{J(k-1)}$$
$$C_p = \frac{Rk}{J(k-1)} \tag{3.46}$$

For the isentropic process we can also readily show that

$$\frac{T_1}{T_2} = \left(\frac{v_2}{v_1}\right)^{k-1} \tag{3.47a}$$

or

$$\frac{T_1}{T_2} = \left(\frac{p_1}{p_2}\right)^{(k-1)/k} \tag{3.47b}$$

The corresponding terms for the polytropic process are,

$$\frac{T_1}{T_2} = \left(\frac{v_2}{v_1}\right)^{n-1} \tag{3.48a}$$

or

$$\frac{T_1}{T_2} = \left(\frac{p_1}{p_2}\right)^{(n-1)/n} \tag{3.48b}$$

Table 3.1 gives the properties of various gases at low pressures and $80°F$.

Table 3.1† Properties of Gases at Low Pressures and 80°F

Gas	Chemical Formula	Molecular Weight	Gas Constant R (ft lb/lb_m °R)	Specific Heat (Btu/lb_m °R)		Specific Heat Ratio k
				C_p	C_v	
Air	—	29.0	53.3	0.240	0.171	1.40
Carbon monoxide	CO	28.0	55.2	0.249	0.178	1.40
Helium	He	4.00	386	1.25	0.753	1.66
Hydrogen	H_2	2.02	766	3.43	2.44	1.40
Nitrogen	N_2	28.0	55.2	0.248	0.177	1.40
Oxygen	O_2	32.0	48.3	0.219	0.157	1.40
Water vapor	H_2O	18.0	85.8	0.445	0.335	1.33

†Reproduced with permission from *Fluid Mechanics* by V. L. Streeter, McGraw-Hill Book Company, Inc., New York, 1962, p. 535.

3.10 Closure

The rational study of the fluid mechanics of steady flow develops from Newton's laws combined with the principle of conservation of energy. For further applications to such topics as compressible flow it will be found necessary to utilize the Second Law of Thermodynamics and an appropriate equation of state for the ideal gas. These items have been developed in this chapter. In Chapter 4 extensive use will be made of the Bernoulli equation, which we developed from the energy equation to indicate its limitations.

The principles utilized and the equations developed in this chapter are not difficult for the student. However, it has been the author's experience that unless these equations are applied with due care to their limitations, with the proper units, and above all with an understanding of their development, they will invariably present great difficulty to the student. Also, since subsequent developments in this text require the use of the material in this chapter, it is very important that it be mastered before proceeding further into the study of fluid mechanics.

REFERENCES

1. *Elementary Applied Thermodynamics* by I. Granet, John Wiley & Sons, Inc., New York, 1965.
2. *Thermodynamics* by E. Fermi, Dover Publications, Inc., New York, 1956.
3. *Heat and Thermodynamics* by M. W. Zemansky, 4th ed., McGraw-Hill Book Company, Inc., New York, 1957.
4. *Concepts of Thermodynamics* by E. F. Obert, McGraw-Hill Book Company, Inc., New York, 1960.
5. *Principles of Engineering Thermodynamics* by P. J. Kiefer, G. F. Kinney, and M. C. Stuart, 2nd ed., John Wiley & Sons, Inc., New York, 1954.
6. *Engineering Thermodynamics* by D. B. Spalding and E. H. Cole, McGraw-Hill Book Company, Inc., New York, 1959.
7. *Chemical Engineering Thermodynamics* by B. F. Dodge, McGraw-Hill Book Company, Inc., New York, 1944.
8. *Fluid Mechanics* by R. C. Binder, 4th ed. Prentice-Hall, Inc., Englewood Cliffs, N. J., 1962.
9. *Elementary Theoretical Fluid Mechanics* by K. Brenkert Jr., John Wiley & Sons, Inc., New York, 1960.
10. "Relation of Force, Acceleration, and Momentum for a Body Whose Mass Is Varying" by H. W. Sibert, *Journal of the Aeronautical Sciences*, 20, February 1953, p. 141.

PROBLEMS

3.1 Ten cubic feet of fluid per second flow in a pipe 1 in. in diameter. If the fluid is water whose specific weight is 62.4 lb/ft^3 determine the weight rate of flow in pounds per hour.

3.2 A gallon is a volume measure of 231 in^3. Determine the velocity in a pipe 2 in. in diameter when water flows at the rate of 20 gal/min in the pipe.

3.3 When a fluid of constant density flows in a pipe, show that the velocity is inversely proportional to the square of the pipe diameter.

3.4 Show that the kinetic energy of a fluid flowing in a pipe varies inversely as the fourth power of the pipe diameter.

3.5 A large pipe (10 in. in diameter) supplies two branch pipes each 4 in. in diameter. Assuming equal weight flow in each of the branch pipes, determine a relationship between the velocity in the main pipe to the velocity in the branch pipes. Assume that γ is constant.

3.6 The flow in a city water system is being tested by allowing it to flow from a hydrant with an outlet 4 in. in diameter. The outlet is 2.0 ft above the ground, and the issuing stream of water hits the level ground at a distance 9 ft from the outlet. How much water is flowing?

3.7 A body of mass of 10 slugs is placed 10 ft above an arbitrary plane. If the acceleration of gravity is 10 ft/sec², how much work (foot pounds) was done in lifting the body above the plane?

3.8 A body weighing 10 lb is lifted 100 ft. What is its potential energy? What velocity will it possess after falling the 100 ft?

3.9 A body weighing 10 lb is moving with a velocity of 50 ft/sec. From what height would it have to fall to achieve this velocity? What is its kinetic energy?

3.10 Water flows over the top of a dam and falls freely until it reaches the bottom some 600 ft below. What is the velocity of the water just before it hits the bottom? What is its kinetic energy per pound at this point?

3.11 A water pump is placed at the bottom of a well. If it is to pump 10 ft³ sec to the surface, 100 ft away, through a 4-in. i.d. pipe, determine the power required in foot pounds per second.

3.12 A hose is 4 in. in diameter and has water flowing in it. If 10 ft³/sec are flowing, determine the velocity in the hose. If the pressure in the pipe is 20 psig, determine the maximum height the water can reach.

3.13 A water turbine operates from a water supply that is 100 ft above the turbine inlet. It discharges to the atmosphere through a 6-in.-diameter pipe with a velocity of 25 ft/sec. If the reservoir is infinite in size, determine the work out of the turbine.

3.14 In a constant pressure process, steam at 500 psia is presented to the piston of a pump and causes the piston to travel 4 in. If the cross-sectional area of the piston is 5 in.², how much work was done by the steam on the piston?

3.15 Air is compressed steadily until its final volume is half its initial volume. The initial pressure is 35 psia and the final pressure is 100 psia. If the initial specific volume is 1 ft³/lb, determine the difference in the p/γ terms at entrance and exit in foot pounds per pound.

3.16 The specific heat of air at constant pressure is 0.24 Btu/lb°F. If 1000 Btu are added to 20 lb of air in a steady flow constant pressure process, what is the temperature increase of the gas?

3.17 Determine the specific heat of an adiabatic process. Is this the constant volume or constant pressure specific heat?

3.18 Air expands through a nozzle from 1000 psia and 500°F to 700 psia and 50°F. If the specific heat of air at constant pressure is 0.24 Btu/lb°F, what is the final velocity? Assume that the initial velocity is zero.

3.19 What is the weight of air in a container of 5 ft³ if it is at 75°F and 100 psia.

3.20 If an ideal gas is made to undergo a process whereby its absolute pressure is doubled and its absolute temperature increases 50 percent, what is the ratio of its final specific volume to its initial specific volume?

NOMENCLATURE

A	= area	v	= specific volume
a	= acceleration	W	= weight
C	= specific heat	W	= work
C_p	= specific heat at constant pressure	$\dot{W}$	= weight rate of flow
C_v	= specific heat at constant volume	x	= distance (length)
F	= force	Z	= height
g	= local acceleration of gravity		
h	= enthalpy	γ	= specific weight
J	= mechanical equivalent of heat	Δ	= small increment
K. E.	= kinetic energy	θ	= angle
k	= C_p/C_v	ρ	= density
lim	= limit		
m	= mass	*Subscripts*	
n	= polytropic exponent		
P. E.	= potential energy	A	= air
p	= pressure	ext	= external
q	= heat	f	= fuel
R	= gas constant	i	= in
R	= reaction force	in, out	= in, out
s	= entropy (specific)	jet	= jet
T	= temperature absolute	o	= out
t	= time	p	= pressure
U	= internal energy	res	= resultant
u	= specific internal energy	v	= volume
V	= velocity	1, 2	= states
$\bar{V}$	= average velocity		

Chapter 4

FLUID DYNAMIC
APPLICATIONS

4.1 Introduction

In Chapter 3 the energy equation, the Bernoulli equation, and momentum relations were derived and their limitations were discussed. It is interesting to note that the Bernoulli equation was first proposed by Daniel Bernoulli in 1738 and since that time has been used extensively to solve many problems in fluid mechanics. The application of Bernoulli's equation often requires that it be modified to account for deviations from the ideal conditions to which it is strictly applicable and usually the equation (or its modifications) yield results that are within the accuracy required for engineering applications. However, there are several cases where the application of this equation will yield incorrect results since it is being used beyond the limits that even its modified forms can be hoped to apply. Several examples of this situation are the high speed flow of compressible fluids, fluid flow with heat transfer causing large density changes, and fluid flow with large pressure changes due to frictional effects. For these cases it is necessary to utilize the continuity equations, the

98

energy equations, the momentum equations, and Newton's laws to obtain analytical solutions. For complex flow problems it may also be necessary to resort to experiments to obtain sufficient information to be able to correlate and solve some of these complex fluid mechanics problems.

This chapter is devoted to the study of those one-dimensional steady flow situations to which the Bernoulli equation and momentum relations can be applied to yield reasonable solutions. Chapter 7, dealing with compressible flow, requires the use of the energy equation due to the large density changes that occur in the fluid. Many practical cases will be studied and analyzed in this chapter, and their limitations will be noted for each case. Usually a flow situation will be noted for each case. Usually a flow situation will be simplified and idealized, and due to this a word of caution must be emphasized at this point. It is quite easy for the student to extend these idealizations to cases where they do not apply. It is necessary to be aware of the assumptions being made, and care must be exercised not to attempt to extend these simplified analyses to complex flow situations for which they are not intended.

4.2 General Considerations

When the energy equation was considered in Chapter 3, it was very often convenient to select a control volume whose control surface coincided with the physical surface of a system. However, this is not necessary, and frequently we shall wish to utilize a control volume whose control surface is selected for ease of analysis. Any fluid stream consists of molecules of the fluid having a random motion due to collisions with other molecules and in addition having a directed motion along the flow path. A streamline is defined as a line with the tangent to it at any point being the direction of its velocity at that point. Streamlines cannot cross since this would require two molecules having different velocities to be at the same point at the same time, and we also note that in steady flow, streamlines do not change with time. Since the streamline coincides with the macroscopic path of the flow, it must be parallel to the surrounding flow and no flow can cross a streamline. If the flow is accelerating, the streamlines are not parallel to each other and will appear closer together, and if the flow is decelerating, the streamlines will appear further apart. For the case of uniform steady flow the streamlines are parallel. These concepts are apparent when the ideal airfoil shown in Figure 4.1 is considered.[1] In the undisturbed stream before and after the airfoil the streamlines are uniformly spaced and parallel. Above

---

[1]See Chapter 9 for a further discussion of flow about an airfoil.

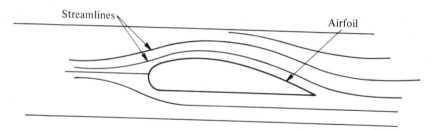

Figure 4.1. Flow over an airfoil.

the airfoil they are close together, indicating a high-velocity region, and beneath the airfoil they are farther apart, indicating a low-velocity region.

As noted, in steady flow no fluid crosses a streamline, and the flow follows the streamlines. Every streamline is a continuous line that may be considered as starting at an infinite distance upstream and extending to a infinite distance downstream. The streamlines through the points of a closed curve will result in a closed surface being generated that is known as a stream tube. Since the

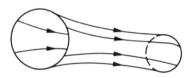

Figure 4.2. Stream tube for steady flow.

surface of the stream tube (Figure 4.2) consists of streamlines, no flow can enter or leave the stream tube through its lateral surface. When a control volume is selected in which the flow is confined by solid walls, streamlines can be considered to start in a reservoir where the fluid is at rest and to extend to the discharge of the system.

The Bernoulli equation derived in Chapter 3 was

$$\frac{p_1}{\gamma_1} + \frac{V_1^2}{2g} + Z_1 = \frac{p_2}{\gamma_2} + \frac{V_2^2}{2g} + Z_2 \tag{4.1}$$

and since the flow was assumed to be incompressible, γ_1 can be taken equal to γ_2 or both can simply be written as γ. Equation (4.1) can also be written as

$$\frac{p_1}{\gamma} + \frac{V_1^2}{2g} + Z_1 = \text{Constant} \tag{4.2}$$

When written in this form we state that the sum of flow energy, the kinetic energy, and the potential energy is a constant *along a streamline*. For another streamline the constant can have a different value, but in many problems all of the streamlines have practically the same total energy. For example, a reservoir that has all of the streamlines starting in it will have the same constant for all of the streamlines. Thus for these cases the sum of the energies on the left side of equation (4.1) or (4.2) can be equated to the sum of the corresponding terms between positions in an ideal system regardless of which streamline is being considered.

All real fluid flow situations are irreversible due to viscous effects giving rise to shear stresses in the fluid. Theoretically the Bernoulli equation as we have derived it must be modified in order to apply it to those cases in which nonideal effects (friction, turbulence, etc.) are relatively large when compared to the constant of equation (4.2). Let us assume that point 1 is upstream and that point 2 is downstream along a streamline. Assuming that there is no addition of energy into or out of the streamline as either work or heat, we can state that the energy at point 1 equals the energy at point 2 plus all of the flow losses between these two points. Mathematically,

$$\frac{p_1}{\gamma_1} + \frac{V_1^2}{2g} + Z_1 = \frac{p_2}{\gamma_2} + \frac{V_2^2}{2g} + Z_2 + \text{Losses}_{1\to2} \qquad (4.3)$$

The loss term on the right-hand side of equation (4.3) can be considered to be a term required to ensure energy conservation. If a pump adds energy to the streamline between points 1 and 2 or a turbine extracts energy between these points, the Bernoulli equation is usually further modified to account for these energy additions or depletions. Thus for a pump,

$$\frac{p_1}{\gamma_1} + \frac{V_1^2}{2g} + Z_1 + E_p = \frac{p_2}{\gamma_2} + \frac{V_2^2}{2g} + Z_2 + \text{Losses}_{1\to2} \qquad (4.4)$$

For a turbine,

$$\frac{p_1}{\gamma_1} + \frac{V_1^2}{2g} + Z_1 = \frac{p_2}{\gamma_2} + \frac{V_2^2}{2g} + Z_2 + \text{Losses}_{1\to2} + E_T \qquad (4.5)$$

For both a pump and a turbine, we have

$$\frac{p_1}{\gamma_1} + \frac{V_1^2}{2g} + Z_1 + E_p = \frac{p_2}{\gamma_2} + \frac{V_2^2}{2g} + Z_2 + \text{Losses}_{1\to2} + E_T \qquad (4.6)$$

The student will note that the modifications to the Bernoulli equation are an attempt to meet the requirement of energy conservation and that the resultant equations bear a resemblance to the energy equation derived in Chapter 3. It should be emphasized that for flows in which there are large density changes due to temperature changes or large additions of depletions of energy from a system causing large density changes, the Bernoulli equation and the energy equation are not equivalent and the apparent resemblance of equations (4.4) through (4.6) to the energy equation should not lead the student to believe that they are applicable to such situations.

In equations (4.3) through (4.6) the term p/γ is referred to as the pressure head, the term $V^2/2g$ is referred to as the velocity head, and the term Z is referred to as the potential head. Using the usual engineering units it will be found that each of these terms has the units of feet, and their sum is called the total head or total energy. These terms are more readily understood by referring to Figure 4.3 on which each term is shown for sections along an irregular tube. In Figure 4.3a the total head or total energy is a constant for flow without friction and is represented by the total head or total energy line as a horizontal line at a constant distance from the datum plane. With friction this is not true and the total head line in Figure 4.3b is shown as a dashed curved line dropping off to the right. It will be noticed that when the velocity in the tube increases, the sum of the potential and pressure heads must decrease, and that for a decreased velocity, the sum of the potential and pressure heads increases. For steady flow, $V_3^2/2g$ is a constant, requiring p_3/γ to be constant for the case of no friction (Figure 4.3a); in the case of friction (Figure 4.3b), p_3/γ must decrease since $V_3^2/2g$ is constant.

ILLUSTRATIVE PROBLEM 4.1. Water is to be delivered from an open tank through a pipeline to a lower elevation. Flow rate is to be 100 gal/min of 60°F water ($\gamma = 62.4$ lb/ft³), and the inside diameter of the pipeline is 2 in. At the pipe outlet the desired pressure is to be 4 psi above atmospheric pressure. If the friction drop in the pipeline is estimated to be 38 ft, determine the vertical distance required between the level in the supply tank and the point of water discharge.

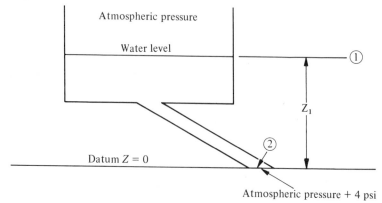

Solution. Referring to the accompanying figure, we shall select two stations, one at the water-air free surface in the tank and the other at the outlet of the pipe. Also, we shall select the potential head datum at the pipe outlet

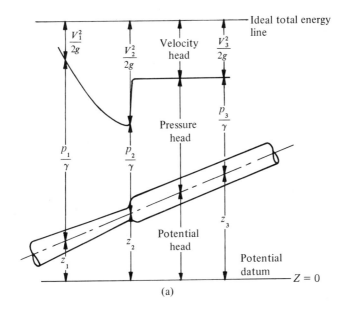

(a)

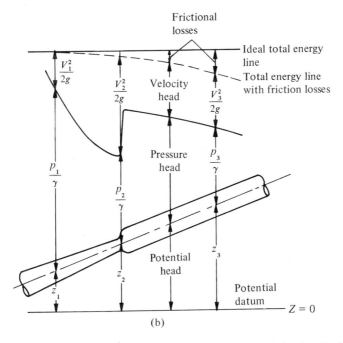

(b)

Figure 4.3. *Head variation in a tube. Adapted with permission from* Basic Fluid Mechanics *by J. L. Robinson, McGraw-Hill Book Company, Inc., New York, 1963, p. 49.*

as shown. For the conditions of this problem we note that all streamlines have the same total energy at the water level in the tank and that therefore we can properly apply the Bernoulli equation at all positions in the flow regardless of the streamline involved. Using equation (4.3),

$$\frac{p_1}{\gamma} + \frac{V_1^2}{2g} + Z_1 = \frac{p_2}{\gamma} + \frac{V_2^2}{2g} + Z_2 + \text{Losses}_{1\rightarrow2}$$

The selection of the datum as shown automatically makes $Z_2 = 0$, and it will be assumed that V_1 is very small compared to V_2 since the tank is presumably large compared to the pipe. Thus for this problem we have

$$\frac{p_1}{\gamma} + Z_1 = \frac{p_2}{\gamma} + \frac{V_2^2}{2g} + \text{Losses}_{1\rightarrow2}$$

However, p_2 exceeds p_1 by 4 psi. Since a column of 1 ft of water of $\gamma = 62.4$ exerts a static pressure at its base equal to $62.4/144$ psi (or 0.433 psi), 4 psi = $4/0.433 = 9.23$ ft. Therefore $(p_2/\gamma) - (p_1/\gamma) = 9.23$ ft:

$$\therefore Z_1 = \left(\frac{p_2}{\gamma} - \frac{p_1}{\gamma}\right) + \frac{V_2^2}{2g} + \text{Losses}_{1\rightarrow2} = 9.23 + \frac{V_2^2}{2g} + 38$$

By definition a gallon is a volume measure equal to 231 in³. Therefore

$$100 \text{ gal/min} = \frac{100}{60} \text{ gal/sec} = \frac{100 \times 231}{60 \times 1728} \text{ feet}^3/\text{sec}$$

Since volume flow equals

$$\frac{100 \times 231}{60 \times 1728} = \left[\frac{\frac{\pi}{4}(2)^2}{144}\right] V_2 \quad \text{and} \quad V_2 = 10.25 \text{ ft/sec}$$

Using this value of V_2,

$$Z_1 = 9.23 + \frac{(10.25)^2}{2 \times g} + 38 = 48.86 \text{ ft} \quad \text{say 49 ft}$$

4.3 Torricelli's Theorem

Let us assume that a very large (essentially infinite area) reservoir has an outlet in its side wall consisting of a small, well-rounded circular opening placed at a distance h below the level of the water in the reservoir. Assuming atmospheric pressure at the water level in the tank and that the jet efflux is to atmospheric pressure, we desire to evaluate the velocity of the water as it issues from the tank. Referring to Figure 4.4, we shall select as the datum for the potential head the level of the water at the center line of the opening. Also, as was noted in illustrative problem 4.1, the total energy content of all of the streamlines in the reservoir is the same and we can select two sections

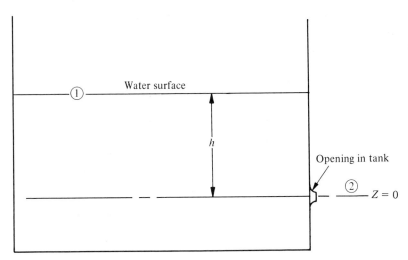

Figure 4.4. *Efflux from a large tank.*

in the system at which we shall apply the Bernoulli equation even though the equation is not being applied to the same streamline. Using the conditions of the problem, noting that the density is constant, and neglecting losses, we have

$$\frac{V_1^2}{2g} + Z_1 + \frac{p_1}{\gamma} = \frac{V_2^2}{2g} + Z_2 + \frac{p_2}{\gamma}$$

Since $Z_2 - Z_1 = h$, and the pressure in the jet must be atmospheric pressure making $p_1/\gamma = p_2/\gamma$,

$$h = \frac{V_2^2}{2g}$$

or

$$V_2 = \sqrt{2gh} \tag{4.7}$$

Equation (4.7) states simply that the velocity of the water from the jet would equal the velocity of a body in free fall from the surface of the reservoir to the center line of the jet; this result is known as Torricelli's theorem.

In deriving equation (4.7) the assumption was made that there were no losses in the flow as it issues from the tank. Actually, for even well-rounded openings, the streamlines will start to converge toward the opening while still in the tank and will continue to converge after leaving the tank. In effect there is a contraction of the stream after leaving the plane of the opening that gives rise to a section of the stream having an area less than the geometric area of the opening in the tank. For a sharp-edged orifice the streamlines are shown as in

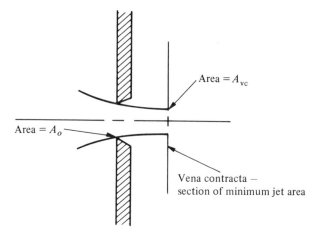

Figure 4.5. Sharp-edged orifice.

Figure 4.5, and the area of minimum cross section is denoted as the vena contracta. By making the edges of the opening rounded or by using a tubular insert into the tank, it is possible to cause a rearrangement of the streamlines to yield a larger diameter at the vena contracta than for the case of a sharp-edged orifice. It is usual in practice to modify equation (4.7) to account for these effects by defining a coefficient of contraction, which is the ratio of the area of the vena contracta to the area of the orifice (or opening). Thus

$$C_C = \frac{A_{vc}}{A_o} \tag{4.8}$$

and we also define the ratio of the actual velocity to the velocity given by equation (4.7) as the coefficient of velocity. This gives

$$V_2 = C_V\sqrt{2gh} \tag{4.9}$$

where C_V is the coefficient of velocity. Since the volume discharging through the orifice or opening is the product of the actual velocity in the jet multiplied by the jet area, we arrive at a new coefficient, called the coefficient of discharge, which is the ratio of the volume actually flowing in the jet to the volume calculated using the area of the opening and the ideal velocity given by equation (4.7). This also gives us the following relation between these three coefficients:

$$C_D = C_C C_V \tag{4.10}$$

For sharp-edged orifices some typical values of these coefficients are $C_c = 0.61$,

$C_V = 0.98$, and $C_D = 0.60$. For well-rounded openings, C_C can go to 1, which makes C_D go toward unity since C_V usually lies in the range of 0.95 to 0.98.

ILLUSTRATIVE PROBLEM 4.2. A large tank has a well-rounded circular opening in its side located 10 ft below the water level in the tank. If the opening is 1 in. in diameter, determine the velocity of the efflux, the area of the vena contracta, and the volume rate of efflux if $C_C = 0.85$ and $C_V = 0.98$.

Solution. From equation (4.7),

$$V_2 = \sqrt{2gh} = \sqrt{2 \times 32.2 \times 10} = 25.4 \text{ ft/sec}$$

The actual velocity is $C_V V_2 = 0.98(25.4) = 24.9$ ft/sec. The area of the opening is $(\pi/4)[(1)^2/144] = 0.00546$ ft^2. The area of the vena contracta is $C_c A_o = 0.85 (0.00546) = 0.00464$ ft^2. We can obtain the volume rate of efflux as the product of the actual velocity multiplied by the area of the vena contracta or as C_D multiplied by the product of the area of the opening and the ideal jet velocity. Thus

$$Q = (24.9) \times (0.00464) = 0.115 \text{ ft}^3/\text{sec}$$

and as a check,

$$Q = (A_o V_2)C_C C_v = (25.4)(0.00546)(0.85)(0.98) = 0.116 \text{ ft}^3/\text{sec}$$

In terms of gallons per minute we have

$$Q = 0.116 \times \frac{1728}{231} \times 60 = 52 \text{ gal/min}$$

since 1 gallon is defined as 231 in^3.

Orifices are frequently used in pipelines to meter the quantity of fluid flowing. When used in this manner the orifice usually consists of a thin plate inserted into a pipe and clamped between flanges. The hole in the orifice plate is concentric with the pipe and static pressure taps are provided upstream and downstream of the orifice, as shown in Figure 4.6. The orifice meter causes a constriction in the flow and creates a jet smaller than itself. Applying the Bernoulli equation to this situation between sections 1 and 2,

$$V_2 = \left[\frac{C_v}{\sqrt{1 - C_c^2(A/A_1)^2}} \right]\left[\sqrt{\frac{2g(p_1 - p_2)}{\gamma}} \right] \qquad (4.11)$$

and

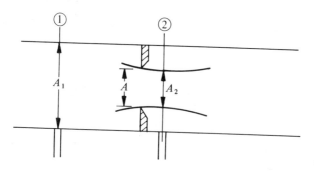

Figure 4.6. *Orifice meter.*

$$Q = AV = \left[\frac{AC_C C_V}{\sqrt{1 - C_C^2 (A/A_1)^2}} \right] \left[\sqrt{\frac{2g(p_1 - p_2)}{\gamma}} \right] \qquad (4.12)$$

From a practical standpoint it is almost impossible to determine C_C and C_V separately for the orifice meter, and it is equally difficult to locate the pressure tap at the vena contracta. Therefore, the simplification is frequently made to combine these coefficients and to rewrite equation (4.12) as

$$Q = AC \sqrt{\frac{2g(p_1 - p_2)}{\gamma}} \qquad (4.13)$$

where C is defined as

$$\frac{C_D}{\sqrt{1 - \left(\dfrac{d_0}{d_1}\right)^4}}$$

and therefore includes the correction for the contraction of the jet. Figure 4.7 shows the coefficient C for a square-edged orifice as a function of the Reynolds number $d_1 V_1 \gamma / \rho g$. It will be noted that the coefficient becomes constant above certain values of the Reynolds number. The advantages of the orifice meter are its relatively small size, the ease of installation in a pipe, and the fact that "standard" installations can be used without the need for calibration. However, the orifice meter acts as a partially open valve and has a low discharge coefficient and consequently a relatively high head loss. In addition, errors can be caused due to poor placement leading to the orifice being eccentric to the inside diameter of the pipe, inaccurate location of the pressure taps, and a rough edge on the orifice opening.

Some of the problems associated with the orifice meter can be alleviated by using a nozzle, as shown in Figure 4.8. The coefficient C is defined as for the orifice meter, and it will be seen that the high value of this coefficient indicates relatively low losses in the nozzle. The manner in which C is defined

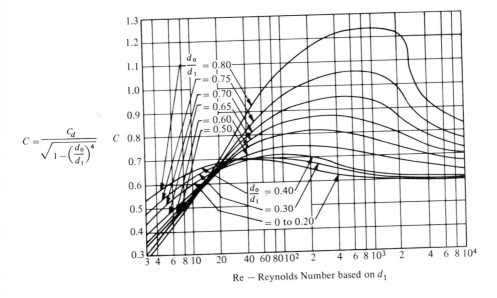

$$C = \frac{C_d}{\sqrt{1 - \left(\frac{d_0}{d_1}\right)^4}}$$

Re — Reynolds Number based on d_1

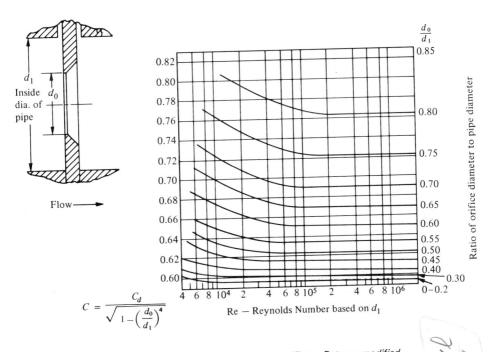

$$C = \frac{C_d}{\sqrt{1 - \left(\frac{d_0}{d_1}\right)^4}}$$

Re — Reynolds Number based on d_1

Figure 4.7. *Flow coefficient* C *for square-edged orifices. Data as modified for* Technical Paper #410, *Crane Co., Chicago, 1957, with permission. Lower-chart data from* Regeln fuer die Durchflussmessung mit genormtem Duesen und Blenden. *VD1-Verlag G.m.b.H., Berlin, SNW,* **7,** *1937. Published as Technical Memorandum 952 by the NACA.*

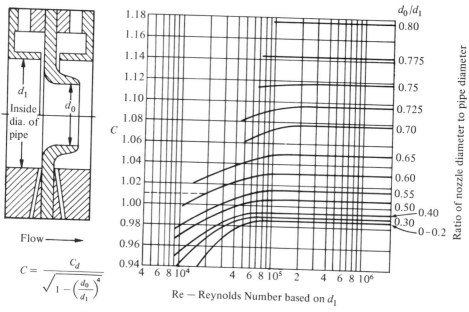

Figure 4.8. Flow coefficient C *for nozzles. Data as modified for* Technical Paper #410, *Crane Co., Chicago. 1957, with permission. Data from* Regeln fuer die Durchflussmessung mit genormtem Duesen und Blenden, *VD1-Verlag G.m.b.H., Berlin, SNW,* **7,** *1937. Published as* Technical Memorandum 952 *by the NACA.*

as well as the location of the pressure taps for this flow nozzle leads to values of C that can exceed unity.

The measurement of the three orifice coefficients (either singly or deriving the third from two of the others) presents an interesting problem in experimental fluid mechanics. We can readily visualize weighing the efflux for a given period of time and thus obtaining the actual rate of discharge. Knowing the area of the opening and the height of fluid above the opening we can directly compute C_D from its definition. We can (in principle) also measure the area of the vena contracta by use of a caliper gage if proper care is taken. This measurement yields C_C, and C_V is calculated from the values of C_D and C_C. However, this method of measuring C_C is not very accurate, and rather than using it, let us attempt to measure C_V and to compute C_C. It is possible to measure C_V in the following manner. Let the jet flow and let us measure the position of a point along the trajectory downstream of the vena contracta. We can a accomplish this by using a beam and sharp-edged pointers or else by simply noting the point at which the jet strikes a plate placed at

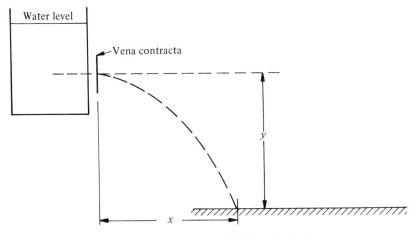

Figure 4.9. *Trajectory method of evaluating C_V.*

some convenient position, as indicated in Figure 4.9. The time for a particle to fall the distance y (neglecting air friction) is given as

$$t = \sqrt{\frac{2y}{g}}$$

and the time for the particle to travel the distance x with the constant velocity of the vena contracta (V_{vc}) is

$$t = \frac{x}{V_{vc}}$$

Therefore

$$V_{vc} = \frac{x}{\sqrt{\frac{2y}{g}}}$$

and

$$C_V = \frac{V_{vc}}{V_2} \tag{4.14}$$

by definition.

ILLUSTRATIVE PROBLEM 4.3. The orifice in illustrative problem 4.2 discharges 65 gal/min under a head of 16 ft. x and y were measured as shown in Figure 4.9 as being, respectively, 17.25 and 5.0 ft. Determine C_C, C_V, and C_D.

Solution. For 16 ft,

$$V_2 = \sqrt{2(g)(16)} = 32.5 \text{ ft/sec}$$

From illustrative problem 4.2, $A_0 = 0.00546$ ft². Therefore the ideal flow is

$$(32.5)(60)(0.00546) \times \frac{1728}{231} = 79.7 \text{ gal/min}.$$

From its definition,

$$C_D = \frac{65}{79.7} = 0.816$$

Using the trajectory data,

$$V_{vc} = \frac{x}{\sqrt{\dfrac{2y}{g}}} = \frac{17.25}{\sqrt{\dfrac{2 \times 5}{g}}} = 30.9 \text{ ft/sec}$$

and

$$C_V = \frac{30.9}{32.5} = 0.952$$

Having C_D and C_V,

$$C_C = \frac{C_D}{C_V} = \frac{0.816}{0.952} = 0.858$$

4.4 Siphon

In a siphon, fluid is made to flow above the level of a source by the pressure force exerted by the atmosphere. The maximum height to which it can go (in the absence of flow losses) can be directly evaluted by considering the problem to be that of a static column of fluid where pressure due to the weight of the column plus the vapor pressure of the fluid must be balanced by the pressure of the atmosphere. For water at 60°F the maximum height to which water will flow of its own accord above a source is approximately 33.4 ft at normal atmospheric pressure.

Figure 4.10 shows two flow situations. In Figure 4.10b the pipe is at a higher level than the level of the source, while in Figure 4.10a the level of fluid in the pipe is always below the level of the source. If a Bernoulli equa-

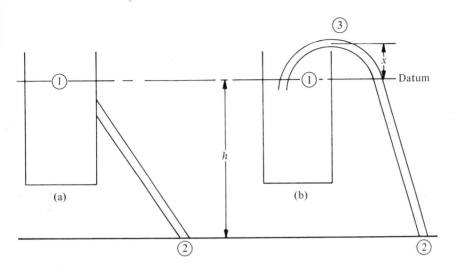

Figure 4.10. Siphon.

tion is written between points 1 and 2 for both of these cases, we obtain two identical equations in the absence of flow losses, yet it is known that if x is made high enough (in the limiting ideal case approximately 33.4 ft for 60°F water), there will be no flow in the arrangement shown in Figure 4.10b. Thus the height x will determine the flow that will exist in the siphon and must be investigated. Let us assume that there are flow losses in the siphon and write the Bernoulli equation between sections $1 \to 3$ and $3 \to 2$. Proceeding we have

$$\frac{V_1^2}{2g} + Z_1 + \frac{p_1}{\gamma} = \frac{V_3^2}{2g} + Z_3 + \frac{p_3}{\gamma} + \text{Losses}_{1\to3} \qquad (4.15)$$

and

$$\frac{V_3^2}{2g} + Z_3 + \frac{p_3}{\gamma} = \frac{V_2^2}{2g} + Z_2 + \frac{p_2}{\gamma} + \text{Losses}_{3\to2} \qquad (4.16)$$

Simplifying these expressions and noting that the pressure at 1 and 2 is atmospheric, p_a,

$$0 + 0 + \frac{p_a}{\gamma} = \frac{V_3^2}{2g} + x + \frac{p_3}{\gamma} + \text{Losses}_{1\to3} \qquad (4.17)$$

and

$$\frac{V_3^2}{2g} + x + \frac{p_3}{\gamma} = \frac{V_2^2}{2g} - h + \frac{p_a}{\gamma} + \text{Losses}_{3\to2} \qquad (4.18)$$

Substituting from (4.18) into (4.17),

$$\frac{p_a}{\gamma} = \frac{V_2^2}{2g} - h + \frac{p_a}{\gamma} + \text{Losses}_{1\to3} + \text{Losses}_{3\to2} \qquad (4.19)$$

which yields

$$\frac{V_2^2}{2g} = h - \text{Losses}_{1\to2} \qquad (4.20)$$

which we obtain by applying the Bernoulli equation directly between sections 1 and 2. Returning to equations (4.17) and (4.18), let us obtain a solution for x, the height of the siphon:

$$x = \frac{p_a}{\gamma} - \frac{V_3^2}{2g} - \frac{p_3}{\gamma} - \text{Losses}_{1\to3} \qquad (4.21)$$

or

$$x = \frac{V_2^2}{2g} - \frac{V_3^2}{2g} - h + \frac{p_a}{\gamma} - \frac{p_3}{\gamma} + \text{Losses}_{3\to2} \qquad (4.22a)$$

If the siphon is flowing full and is of constant cross section, $V_2 = V_3$ and equation (4.22a) becomes

$$x = \frac{p_a}{\gamma} - \frac{p_3}{\gamma} - h + \text{Losses}_{3\to2} \qquad (4.22b)$$

From equations (4.20) through (4.22b) the siphon can be analyzed. This is best shown in illustrative problem 4.4.

ILLUSTRATIVE PROBLEM 4.4. A siphon is operated to discharge water at the rate of 1 ft³/sec from a reservoir through a pipe whose inside diameter is 4 in. If the height of the siphon is 10 ft above the level in the reservoir and the discharge of the siphon is 2.02 ft below the reservoir level, determine the velocity in the pipe and the pressure at the highest point in the pipe if (a) flow losses are negligible; (b) if the losses from 1 → 3 are equivalent to a head of 0.1 ft of water and represent one third of the total losses in the pipe, determine the quantity of fluid flowing, the velocity in the pipe, and the pressure at the highest point in the pipe.

Solution. a. Refer to Figure 4.10b. The quantity of fluid flowing is given as 1 ft³/sec. From continuity,

$$V = \frac{Q}{A} = \frac{1}{\dfrac{\pi}{4} \dfrac{(4)^2}{144}} = 11.4 \text{ ft/sec}$$

and using equation (4.21) with no flow losses,

$$\frac{p_3}{\gamma} = \frac{p_a}{\gamma} - \frac{V_3^2}{2g} - x = \frac{14.7 \times 144}{62.4} - 10 = 21.98 \text{ ft}$$

or $34 - 21.98 = 12.02$ ft of vacuum at the summit. For the case of no flow losses we note that this is equal to the head difference between the summit of the siphon to its exit; that is, $10 + 2.02 = 12.02$ ft.

b. In this part of the problem the quantity of fluid flowing is not known. We therefore apply equation (4.20) and from the statement of the problem note that losses $1 \rightarrow 2 = 0.3$ ft. Thus

$$\frac{V_2^2}{2g} = h - 0.3 = 2.02 - 0.3$$

and

$$V_2 = 10.5 \text{ ft/sec}$$

The quantity flowing is therefore

$$\frac{\pi}{4} \times \frac{(4)^2}{144} \times 10.5 = 0.918 \text{ ft}^3/\text{sec}$$

Once again applying equation (4.21),

$$\frac{p_3}{\gamma} = \frac{p_a}{\gamma} - \frac{V_3^2}{2g} - x - \text{Losses}_{1 \to 3}$$

$$= \frac{14.7 \times 144}{62.4} - \frac{(10.5)^2}{2g} - 10 - 0.1$$

$$= 21.88 \text{ ft}$$

and this is $34 - 21.88$ ft, or 12.12 ft, vacuum at the summit.

When analyzing siphons it is important to verify that the summit pressure p_3 is not impossible and also to check that the pipe is flowing full by independently checking the summit velocity against the discharge that would be obtained if the siphon is operated at its maximum height of atmospheric pressure minus the vapor pressure of the fluid (the pressure exerted by vapor in equilibrium with its fluid at the temperature of the fluid).

4.5 Fan or Propeller

The function of a fan or propeller is to transform efficiently its rotary motion into forward linear motion of an airstream to provide an increased air velocity. The blades of such a device can be considered to be a number of airfoils of very short span set end to end that must absorb the power furnished to the propeller by its driving motor. The theory of the action of these small airfoils is known as the Drzewiecki theory or blade-element theory, and its development will be found in standard texts on aerodynamics or propellers. For the present we shall consider a somewhat simplified treatment in which the fan or propeller will be considered as a disk that exerts a uniform pressure or thrust on the airstream over its cross-sectional area, thereby increasing the momentum of the air without causing any rotation of the fluid. The further assumption will be made that the pressure at stations far enough in front of and in back of the propeller is atmospheric and that the flow is streamline from entrance to exit of the device.

Before proceeding with the analysis let us first recast the Bernoulli equation into a more convenient form. Repeating equation (4.2), we have

$$\frac{p_1}{\gamma} + \frac{V_1^2}{2g} + Z_1 = \text{Constant} \tag{4.2}$$

Multiplying equation (4.2) by γ and noting that $\gamma = \rho g$,

$$p_1 + \frac{V_1^2 \gamma}{2g} + \gamma Z_1 = p_1 + \frac{\rho V_1^2}{2} + \gamma Z_1 = \text{Constant} \tag{4.23}$$

For problems in gas flow and gas dynamics the elevation term is usually negligible when compared to the other terms, which leads to a form of the Bernoulli equation sometimes called the aerodynamic form of the Bernoulli equation, namely,

$$p + \frac{\rho V^2}{2} = \text{Constant} \tag{4.24}$$

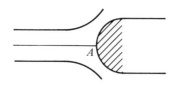

Figure 4.11. Streamlines around a submerged body.

If a solid body is immersed in a fluid such as is shown in Figure 4.11, there will be a point A on the body where the velocity of the fluid is zero. Since the fluid is stagnant at this point, it is known as the stagnation point and the pressure at this point will exceed the pressure of the surrounding stream by $\rho V^2/2$, as determined by equation (4.24).

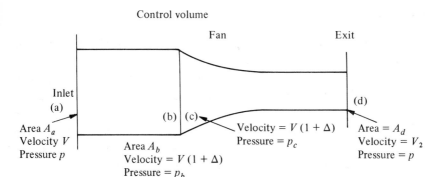

Figure 4.12. *Ideal fan.*

Let us now consider an idealized propeller as shown in Figure 4.12. We shall select four stations along the control volume, namely, the inlet, just prior to the fan, just after the fan, and the outlet. From the continuity equation, which requires the weight of air flowing past each section to be the same, and assuming a negligible change in the density of the air through the fan,

$$A_a V = A_b V(1 + \Delta) = A_d V_2 \tag{4.25}$$

From inlet up to the fan disk no energy enters or leaves the airstream. Therefore,

$$p + \frac{\rho V^2}{2} = p_b + \frac{\rho}{2}\left[V(1 + \Delta)\right]^2 \tag{4.26}$$

Simplifying,

$$p_b = p + \frac{\rho}{2}V^2\left[1 - (1 + \Delta)^2\right] \tag{4.27}$$

Using similar reasoning that no energy enters or leaves the stream from the point at which it leaves the fan to the exit, we have

$$p_c + \frac{\rho}{2}\left[V(1 + \Delta)\right]^2 = p + \frac{\rho}{2}(V_2)^2 \tag{4.28}$$

and

$$p_c = p + \frac{\rho}{2}\{V_2^2 - [V(1 + \Delta)]^2\} \tag{4.29}$$

In this equation (4.29) it has been assumed that the velocity does not change across the section of the fan. If we had assumed a velocity change at this section it would require an infinite acceleration, which is physically impossible. The thrust exerted on the airstream by the fan is equal to the pressure difference across the fan multiplied by the area of the fan. Using equations (4.27) and (4.29), we have

$$T = (p_c - p_b)A_b = A_b\left(p + \frac{\rho}{2}\{V_2^2 - [V(1 + \Delta)]^2\}\right.$$

$$\left. - \left\{p + \frac{\rho}{2}V^2[1 - (1 + \Delta)^2]\right\}\right)$$

and

$$T = A_b\frac{\rho}{2}(V_2^2 - V^2) \tag{4.30}$$

In arriving at equation (4.30) we have essentially utilized the Bernoulli equation as an energy equation. It is also true that the thrust is equal to the rate of change of momentum of the airstream. Mathematically,

$$T = \frac{\dot{w}}{g}(V_2 - V) \tag{4.31}$$

where $V_2 - V$ is the change in velocity of the airstream and $\dot{w}$ is the weight rate of air flow in pounds per second. Using the propeller disk as the reference area,

$$\dot{w} = \gamma A_b V(1 + \Delta) \tag{4.32}$$

Therefore,

$$T = \frac{\gamma A_b V(1 + \Delta)(V_2 - V)}{g} = \rho A_b V(1 + \Delta)(V_2 - V) \tag{4.33}$$

Equating equation (4.30) with equation (4.33), we have

$$A_b\frac{\rho}{2}(V_2^2 - V^2) = \rho A_b V(1 + \Delta)(V_2 - V) \tag{4.34}$$

Simplifying,

$$\frac{V_2 + V}{2} = V(1 + \Delta) \tag{4.35}$$

From equation (4.35) we conclude that the velocity at the disk $[V(1 + \Delta)]$ is the arithmetic mean of the initial and final velocities with half of the total change in velocity occurring before the air passes through the fan and half being imparted to the air after it passes through the fan.

For a fan or propeller we also desire to know how efficient the device is in increasing the kinetic energy of the airstream. The difference in kinetic energy across the unit is

$$K.E._{out} - K.E._{in} = \frac{1}{2}\frac{\dot{w}}{g}(V_2^2 - V^2)$$

$$= \frac{1}{2}\rho A_b V(1 + \Delta)\{V_2^2 - V^2\} \qquad (4.36)$$

The work done on the airstream is the product of the thrust multiplied by the inlet air velocity, namely,

$$\text{Work} = \rho A_b V(1 + \Delta)(V_2 - V)(V) \qquad (4.37)$$

where the thrust is obtained from equation (4.33). Defining the efficiency of this device as the ratio of work divided by the difference in kinetic energy, we have

$$\text{Efficiency } \eta = \frac{\text{Work}}{\text{Change in kinetic energy}} = \frac{\rho A_b V(1 + \Delta)(V_2 - V)(V)}{\frac{1}{2}\rho A_b V(1 + \Delta)(V_2^2 - V^2)}$$

$$(4.38)$$

Using equation (4.35), equation (4.34) simplifies to

$$\eta = \frac{1}{1 + \Delta} \qquad (4.39)$$

Equation (4.39) expresses the efficiency of an ideal fan or propeller in which the only energy loss considered is the loss in kinetic energy of translation. An actual device will have an efficiency less than that given by equation (4.35) due to the fact that the flow is not streamline and that there are friction losses in the blades, pulsations due to the finite number of blades, tip losses, and stream rotation after leaving the blades. Propellers have actual efficiencies of 80–90 percent.

ILLUSTRATIVE PROBLEM 4.5. What is the maximum efficiency of a propeller if the air speed of a plane is 150 mph and the airstream velocity leaving the propeller is 200 mph?

Solution. The increment in velocity across the entire unit is $200 - 150 = 50$ mph. Since half of this is added before the propeller, the velocity at the propeller is $150 + 25 = 175$ mph. Therefore,

$$V(1 + \Delta) = 175$$

$$(1 + \Delta) = \frac{175}{150} = 1.165$$

and

$$\eta = \frac{1}{1 + \Delta} = \frac{1}{1.165} = 85.8 \text{ percent}$$

ILLUSTRATIVE PROBLEM 4.6. An airplane is traveling at 300 mph and has a propeller that is 6 ft in diameter. If the specific weight of air is 0.0755 lb/ft^3, determine the thrust provided by the propeller if it is 90 percent efficient.

Solution. From equation (4.39), $\eta = 1/(1 + \Delta)$, where Δ is the velocity increment fraction added to the stream before the propeller. For 90 percent efficiency, $\Delta = 0.11$. Since 300 mph $= 440$ ft/sec, the velocity added to the stream before the propeller is $0.11 \times 440 = 48.4$ ft/sec. Thus at the propeller the velocity is $440 + 48.4 = 488.4$ ft/sec. The velocity increment added after the propeller equals that which was added before the propeller. Therefore the exit velocity from the propeller disk is $488.4 + 48.4 = 536.8$ ft/sec. From equation (4.33),

$$T = \rho A_b V(1 + \Delta)(V_2 - V)$$

and

$$T = \frac{0.0755}{32.2} \left(\frac{\pi}{4} \times 6^2 \right) \{440(1.11)\}\{536.8 - 440\}$$

$$= 3140 \text{ lb}$$

where T is thrust.

4.6 Pressure and Velocity Measurements[2]

The measurement of any property of a system requires that the method used should not change the magnitude of the property being measured, thereby yielding an erroneous result. There are many methods available to deter-

[2]Appendix B should be consulted in conjunction with this section.

mine the pressure and velocity at a given location in a flow system, and the student is referred to the technical literature for detailed discussions on this subject. In particular the latest publication of the American Society of Mechanical Engineers, *Flow Meters: Their Theory and Application*, and the *Transactions of the ASME* are recommended for this purpose. At present we shall be concerned solely with those devices to which we can apply the Bernoulli equation to evaluate the desired results.

4.6a Piezometer

The measurement of the pressure of a fluid in motion is particularly difficult since the presence of the measuring device will almost invariably alter the flow or change the magnitude of the pressure. An opening in the wall of a pipe is quite often used to measure the pressure of a flowing fluid. This static pressure is presumed to be the pressure of the undisturbed fluid, and the opening is known as a piezometer opening. The reading obtained with a single opening of this type will not be correct if the opening is not normal to the pipe surface or if the edges of the opening have burrs, and this method of pressure measurement is subject to large errors due to the presence of small perturbations in the installation of the piezometer opening (Figure 4.13). To obtain a

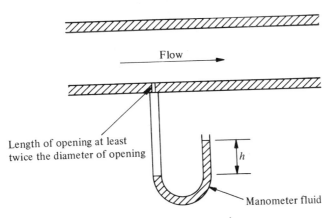

Figure 4.13. Piezometer opening.

more correct reading several holes are often placed around the periphery of a pipe in a given plane and connected together. This device is known as a piezometer ring and the pressure is read directly on a manometer attached to the common pressure connection.

When a body is submerged in a fluid such as in the case of an airfoil, it is often desirable to obtain the static pressure distribution along its boundaries. This can be determined by making piezometer openings along the surface and taking the corresponding static pressure readings by attaching tubes to these openings. In the case of an airfoil we can use these pressure readings to obtain the velocity distribution by applying the Bernoulli equation.

4.6b Static Tube

It is at once apparent that the piezometer opening (or piezometer ring) does not measure the pressure across the section of pipe. Basically it gives the static pressure on a given streamline. Thus the measurement of static pressure on any streamline must be performed with another type of instrumentation. The instrument shown in Figure 4.14 consists of a closed-end tube inserted into the

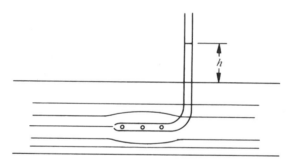

Figure 4.14. Simple static tube.

flow stream at a desired location. This tube, known as the static tube, is made as small as practical so as to minimize any flow disturbances due to its presence, and it contains small radial holes in the leg parallel to the flow. In practice there is some flow disturbance due to the distortion of the streamlines and the leg that is perpendicular to the flow, which can lead to readings either too high or too low. It is possible to calibrate the tube either by placing it in a flow where the static pressure is known from a piezometer reading or by towing it in a still fluid (whose pressure is known) at various velocities. If the difference between the correct static pressure and that read by the static tube is called the static pressure discrepancy (or D), we can state simply that the discrepancy will usually vary as the velocity head; that is,

$$D = C\frac{V^2}{2g} \tag{4.40}$$

where C is an empirical constant determined by plotting D vs. $V^2/2g$.

ILLUSTRATIVE PROBLEM 4.7. A static tube is being calibrated by towing it through a fluid at different velocities. If the measured discrepancy is 0.5 psi when it is towed at 15 ft/sec, what is the discrepancy at 10 ft/sec? Assume that the fluid is water whose specific weight is 62.4 lb/ft³ and calculate the value of C in equation (4.40).

Solution. Using equation (4.40) and assuming C to be a constant,

$$\frac{D_{10}}{D_{15}} = \frac{V_{10}^2}{V_{15}^2}$$

and

$$D_{10} = D_{15}\frac{V_{10}^2}{V_{15}^2} = 0.5\left(\frac{10}{15}\right)^2 = 0.22 \text{ psi}$$

Also,

$$D = C\frac{V^2}{2g}$$

From the data, using consistent units,

$$\frac{0.5(144)}{62.4} = C\frac{(15)^2}{2g}$$

and

$$C = \frac{0.5(144)(2)(32.2)}{(15)^2(62.4)} = 0.33$$

4.6c The Pitot Tube and Pitot-Static Tube

Henri Pitot, a Frenchman, inserted an open-ended tube into the Seine River near Paris in the early 1700's and found that the rise of water in the vertical leg of the tube was proportional to the square of the velocity of the stream at the location at which he made his measurements. Tubes of this type are used to measure the local velocity of a flowing fluid and have been called Pitot tubes in honor of Henri Pitot. Figure 4.15 shows a simple Pitot tube inserted into a stream with its open end facing into the stream flow. Both the tube and its opening are made as small as is practical to keep flow disturbances to a minimum due to the presence of the tube. The assumption is made that the streamline at the center of the Pitot tube is not disturbed by the tube. At the

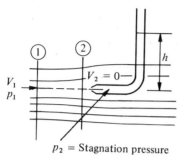

Figure 4.15. *Pitot tube.*

inlet to the tube the stream velocity is reduced to zero and the level of liquid rises in the tube until it reaches the level h above the center line of the tube. It is assumed that the pressure at section 2 is the stagnation pressure of the streamline in question. By applying the Bernoulli equation to the streamline at the center of the tube successively at sections 1 and 2 in the flow, we obtain,

$$\frac{V_1^2}{2g} + \frac{p_1}{\gamma} = \frac{p_2}{\gamma}$$

and

$$V = \sqrt{\frac{2g(p_2 - p_1)}{\gamma}} \tag{4.41}$$

The term $p_2 - p_1$ is sometimes known as the impact, dynamic, or velocity pressure of a stream and it is equal to $\frac{1}{2}\rho V^2$. In equation (4.41) the specific weight, γ, is that of the flowing stream and not that which exists in the manometer leg. If the Pitot tube is installed in an open channel, then $(p_2 - p_1)/\gamma$ is simply h, as shown in Figure 4.16. For a closed conduit it is necessary to

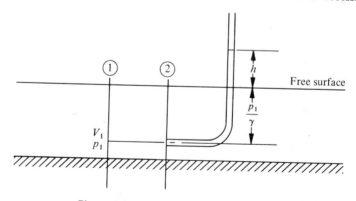

Figure 4.16. *Pitot tube in an open channel.*

evaluate $(p_2 - p_1)/\gamma$ by making a second measurement. This can be done by combining a piezometer tube with the Pitot tube or using a piezometer opening in the pipe wall. Let us assume a piezometer opening is used in conjunction with a Pitot tube, as shown in Figure 4.17, to measure the velocity of a

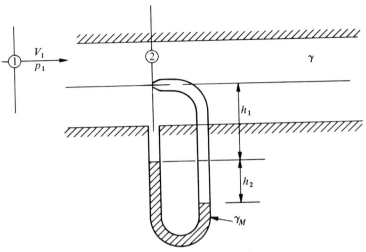

Figure 4.17. *Pitot tube–piezometer for velocity measurement.*

fluid in a pipe. Further assume that a suitable manometer fluid whose specific weight is γ_M will be used to evaluate the pressure differential between the Pitot tube and the piezometer (as shown in Figure 4.17) and let us write the Bernoulli equation between points 1 and 2:

$$\frac{V_1^2}{2g} + \frac{p_1}{\gamma} = \frac{p_2}{\gamma}$$

By evaluation of the manometer we have

$$p_1 + \gamma h_1 + \gamma_M h_2 - \gamma(h_1 + h_2) = p_2$$

Simplifying,

$$\frac{p_2 - p_1}{\gamma} = h_2\left(\frac{\gamma_M}{\gamma} - 1\right) \tag{4.42a}$$

and

$$V_1 = \sqrt{2gh_2\left(\frac{\gamma_M}{\gamma} - 1\right)} \tag{4.42b}$$

In the system shown in Figure 4.17 it is necessary to exercise care in installing the piezometer tap in order to avoid errors in the measurement that have been previously noted.

ILLUSTRATIVE PROBLEM 4.8. Water flows in a pipe. A Pitot-static tube uses mercury as the metering fluid, and a deflection of 2 in. is noted on the manometer. Calculate the flow velocity. $\gamma_W = 62.4$ and $\gamma_M = 850.0$.

Solution. Referring to Figure 4.17 and applying equation (4.42), we have

$$\frac{p_2 - p_1}{\gamma} = h_2\left(\frac{\gamma_M}{\gamma} - 1\right) = \frac{2}{12}\left(\frac{850}{62.4} - 1\right) = 2.1\,\text{ft}$$

$$\therefore \frac{V_1^2}{2g} = 2.1$$

$$V_1 = 11.7\,\text{ft/sec}.$$

The arrangement shown in Figure 4.17 has the disadvantage of requiring two openings in the pipe wall, requires care in making the piezometer opening, and is usually rigid so that only a single measurement point can be obtained. To overcome these objections the Pitot tube is combined with a static tube in a single unit called the Pitot-static tube. It usually consists of two tubes installed one inside of the other and connected to a manometer, as shown in Figure 4.18. The inner tube is open at the end and measures the total pressure of the stream, while the outer tube has holes drilled in it normal to the direction of the flow and measures the static pressure of the stream. With the manometer connections as shown in Figure 4.15, the manometer reads the dynamic pressure. Using an analysis similar to that used for the system shown in Figure 4.17, we obtain equation (4.43) for the system shown in Figure 4.18, namely,

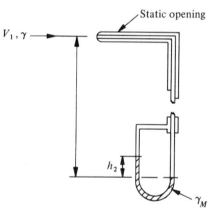

Figure 4.18. *Combined pitot-static tube.*

$$V_1 = \sqrt{2gh_2\left(\frac{\gamma_M}{\gamma} - 1\right)} \tag{4.43}$$

The accuracy of the Pitot-static tube is dependent on its construction and proper installation in the pipe. In addition, it is quite sensitive to misalignment with the flow. Although there are many Pitot-static tubes of standard size, shape, and construction that are intended to minimize these problems, it is always advisable to calibrate a particular installation. When this is done, equation (4.43) is modified to account for the nonideal action of the tube by introducing an empirical coefficient C:

$$V_1 = C\sqrt{2gh_2\left(\frac{\gamma_M}{\gamma} - 1\right)} \qquad (4.44)$$

ILLUSTRATIVE PROBLEM 4.9. A stream of water flows at 10 ft/sec and the manometer (Figure 4.18) indicates a differential of 2 in. Hg. Determine C for this Pitot-static tube. $\gamma_W = 62.4$, $\gamma_M = 850.0$.

Solution. Using equation (4.44),

$$C^2 = \frac{V_1^2}{2gh_2\left(\frac{\gamma_M}{\gamma} - 1\right)} = \frac{10^2}{2(32.2)\left(\frac{2}{12}\right)\left(\frac{850}{62.4} - 1\right)}$$

$$= 0.74$$

and $C = 0.86$

4.6d Venturi Meter

As previously noted, the Pitot tube is used to measure the velocity at a point in a flow stream. For a permanent installation in a pipe that can be used to continuously monitor the rate of fluid with a minimum of flow losses, a venturi meter is frequently utilized. The venturi meter consists of a converging-diverging conical section of pipe arranged to give an increase in velocity and kinetic energy of the stream, thereby causing a measurable drop in pressure in the converging section. The diverging section is used to reconvert the increased kinetic energy of the stream back to pressure energy at the outlet of the meter with a minimum of turbulence and friction loss. The cone angle at the entrance to the throat is usually 20 to 30 deg, while the cone of the exit section is usually from 5 to 14 deg.

In a simplified installation, as shown in Figure 4.19, a manometer is con-

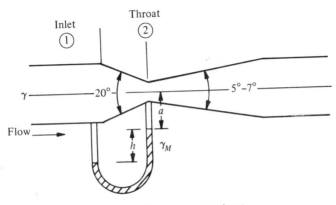

Figure 4.19. Simple venturi meter.

nected to the inlet and throat of the venturi meter. Let us assume that there are no flow losses in the meter and write the Bernoulli equation for sections 1 and 2; we obtain,

$$\frac{p_1}{\gamma} + \frac{V_1^2}{2g} = \frac{p_2}{\gamma} + \frac{V_2^2}{2g}$$

and

$$\frac{V_2^2}{2g} - \frac{V_1^2}{2g} = \frac{p_1}{\gamma} - \frac{p_2}{\gamma} \tag{4.45}$$

The continuity equation requires that for the steady flow of an incompressible fluid $A_1 V_1 = A_2 V_2$. Therefore equation (4.45) becomes

$$\frac{p_1}{\gamma} - \frac{p_2}{\gamma} = \frac{V_2^2}{2g}\left[1 - \left(\frac{A_2}{A_1}\right)^2\right] \tag{4.46}$$

and the volume rate of fluid flow (AV) is

$$AV = A_2 V_2 = \left[\frac{A_2}{\sqrt{1 - \left(\frac{A_2}{A_1}\right)^2}}\right]\sqrt{\frac{2g(p_1 - p_2)}{\gamma}} \tag{4.47}$$

where the specific weight γ is that of the flowing fluid and not that of the manometer fluid shown in Figure 4.19. Using the manometer and a recording instrument it is possible to obtain a continuous record of the flow in a pipe. For the case of a real fluid with friction we introduce a coefficient C_V into equation (4.47). Therefore,

$$(AV)_{\text{actual}} = \frac{C_V A_2}{\sqrt{1 - \left(\frac{A_2}{A_1}\right)^2}}\sqrt{\frac{2g(p_1 - p_2)}{\gamma}} \tag{4.48}$$

In terms of the manometer shown in Figure 4.19,

$$(AV)_{\text{actual}} = \frac{C_v A_2}{\sqrt{1 - \left(\frac{A_2}{A_1}\right)^2}}\sqrt{2g\left(\frac{\gamma_M}{\gamma} - 1\right)} \tag{4.49}$$

The velocity coefficient of a venturi meter is not constant and is found to be a function of the Reynolds number, the area ratio A_2/A_1, and the shape of the venturi. For greatest accuracy an in-place calibration is most desirable. Figure 4.20 shows C_V for venturi meters of different sizes but having a 2 : 1 diameter ratio. In most instances of practical interest C_V can be taken to be 0.9 or greater.

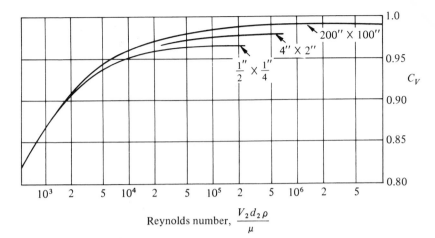

Figure 4.20. *Venturi meter coefficients. Reproduced with permission from* Elementary Fluid Mechanics *by J. K. Vennard, 4th ed., John Wiley & Sons, Inc., New York, 1961, p. 418.*

The venturi meter is a low head loss device since its diverging downstream section past the throat of the venturi acts to recover kinetic energy. The head loss from inlet to throat can be derived from equation (4.3), yielding

$$\text{Losses}_{1\to2} = \left[\left(\frac{1}{C_V^2} - 1\right)\left(1 - \left(\frac{A_2}{A_1}\right)^2\right)\right]\frac{V_2^2}{2g} \tag{4.50}$$

For most installations the term in brackets in equation (4.50) will be from 0.1 to 0.2.

ILLUSTRATIVE PROBLEM 4.10. A venturi meter consists of a 5-in.-diameter inlet and a 2-in.-diameter throat. If h in Figure 4.19 is 10 in. Hg ($\gamma = 850$ lb/ft³), determine the ideal quantity of water flowing.

Solution. From the manometer we have

$$p_1 + \gamma_w(h + b) - \gamma_m h - \gamma_w b = p_2$$

or

$$p_1 - p_2 = h(\gamma_m - \gamma_w)$$

where the subscripts m and w refer to mercury and water, respectively. Thus

$$\frac{p_1 - p_2}{\gamma_w} = \frac{10}{12}\left(\frac{850}{62.4} - \frac{62.4}{62.4}\right) = 10.5 \text{ ft of water}$$

Using equation (4.47),

$$AV = \frac{A_2}{\sqrt{1 - (A_2/A_1)^2}}\sqrt{2g(10.5)}$$

$$\left.\begin{array}{l} A_1 = \dfrac{\pi}{4}\dfrac{(5)^2}{144} = 0.136 \text{ ft}^2 \\[2mm] A_2 = \dfrac{\pi}{4}\dfrac{(2)^2}{144} = 0.0218 \text{ ft}^2 \end{array}\right\} \quad \dfrac{A_2}{A_1} = 0.16$$

$$\therefore AV = \frac{0.0218}{\sqrt{1 - (0.16)^2}}(25.5) = 0.56 \text{ ft}^3/\text{sec}$$

4.7 Deflection of Fluid Streams

When a fluid stream undergoes a change in either its speed or direction, a force must have been exerted on the stream. Concurrently, an equal and opposite force must have been exerted by the fluid stream on the body causing the velocity change. If the deflector is fixed, it is necessary for the structure to be capable of resisting the resultant force exerted on it while the same force exerted on a moving deflector (vane) is capable of doing useful work. The ability of a fluid stream to do work on a moving vane is the basis upon which the theory of turbomachinery is based.

Let us consider the situation of a fixed vane, shown in Figure 4.21, and ap-

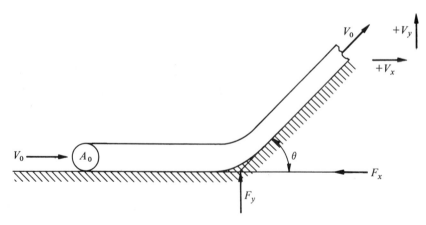

Figure 4.21. Fixed deflector vane.

ply the impulse-momentum relations developed in Chapter 3. The fluid stream of area A_o with an initial horizontal velocity V_o is deflected through the angle θ. Assume the frictional resistance of the stream on the vane to be negligible and the velocity of the stream to be uniform. For a free jet the pressure at both inlet to the vane and outlet of the vane must be equal since they must equal the surrounding (atmospheric) pressure. Therefore, the speed (not velocity) of the jet must be the same at inlet as at outlet; the vane changes only the direction of the stream. The force exerted by the vane on the stream in the horizontal direction is;

$$F_x = \frac{\dot{w}}{g}(V_x - V_o) = \frac{\dot{w}}{g}(V_o \cos\theta - V_o) \qquad (4.51a)$$

and

$$F_x = \frac{-\gamma A V_o^2}{g}(1 - \cos\theta) \qquad (4.51b)$$

In the vertical direction,

$$F_y = \frac{\dot{w}}{g}(V_y - 0) = \frac{\dot{w}}{g}(V_o \sin\theta) \qquad (4.52a)$$

and

$$F_y = \frac{\gamma A V_o^2}{g}\sin\theta \qquad (4.52b)$$

The resultant force

$$R = \sqrt{F_x^2 + F_y^2} = \frac{\gamma A V_o^2}{g}\sqrt{2(1 - \cos\theta)} \qquad (4.53)$$

The derivation of equations (4.51a) through (4.53) has assumed that the difference in elevation between the ends of the vane is negligible. Note, too, that the forces given by these equations are the forces exerted on the fluid stream. The forces on the fixed vane will be equal and opposite to these forces.

ILLUSTRATIVE PROBLEM 4.11. Water is deflected by a 90-deg bend that is in a horizontal plane (both ends at the same elevation). If the stream is 3 in. in diameter, $\gamma = 62.4 \text{ lb/ft}^3$, and the stream velocity is 4 ft/sec, determine the forces exerted on the stream by the vane.

Solution

$$A = \frac{\pi}{4}\left(\frac{1}{4}\right)^2 = 0.0491 \text{ ft}^2$$

$$F_x = -\frac{62.4(0.0491)(4)^2}{32.2}(1 - 0) = -1.52 \text{ lb}$$

$$F_y = \frac{62.4(0.0491)(4)^2}{g}(1) = 1.52\,\text{lb}$$

and

$$R = \sqrt{(-1.52)^2 + (1.52)^2} = 2.16\,\text{lb}$$

If the vane is moving with the velocity u in the same direction as the initial velocity of the stream, we have the situation shown in Figure 4.22. At inlet the

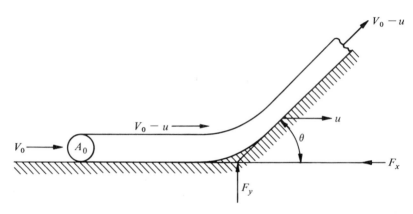

Figure 4.22. *Moving deflector vane.*

absolute velocity of the stream is V_o, and its velocity relative to the vane is $V_o - u$. The relative velocity is turned by the vane through the angle θ without change in magnitude. This relative velocity vector $V_o - u$ is added vectorially to u to give the final absolute velocity leaving the vane. The vector diagram shown in Figure 4.23 shows the situation at the outlet, where V_1 is the resultant velocity (absolute) at the outlet. The velocity component of the absolute velocity in the horizontal direction at the outlet is $(V_o - u)\cos\theta + u$, and at the inlet it is V_o. Therefore,

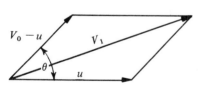

Figure 4.23. *Velocity diagram.*

$$F_x = \frac{\dot{w}}{g}[(V_o - u)\cos\theta + u - V_o] = -\frac{\dot{w}}{g}(V_o - u)(1 - \cos\theta) \quad (4.54)$$

and

$$F_y = \frac{\dot{w}}{g}[(V_o - u)\sin\theta] \quad (4.55)$$

Notice from equations (4.54) and (4.55) that if $V_o - u$ is replaced by the relative velocity, the problem of the moving vane is reduced to the problem of the fixed vane with a velocity equal to the relative velocity.

ILLUSTRATIVE PROBLEM 4.12. If the vane in illustrative problem 4.11 has a velocity of 2 ft/sec to the right, determine the forces acting on the stream.

Solution. $V_o - u$ is $4 - 2 = 2$ ft/sec. Therefore,

$$F_x = -\frac{62.4 \times 0.0491 \times 4}{32.2}[2 - 0] = -0.76\,\text{lb}$$

and

$$F_y = \frac{62.4 \times 0.0491 \times 4}{32.2}[2] = 0.76\,\text{lb}$$

and

$$R = \sqrt{(-0.76)^2 + (0.76)^2} = 1.08\,\text{lb}$$

ILLUSTRATIVE PROBLEM 4.13. If the stream in illustrative problem 4.12 is turned through an angle of 180 deg (a U bend), determine the forces on the stream.

Solution. Refer to the accompanying diagram. Using the absolute velocities from the diagram,

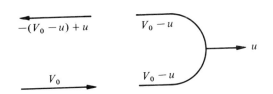

$$F_x = \frac{\dot{w}}{g}[-(V_o - u) + u - V_o] = \frac{\dot{w}}{g}[-2(V_o - u)]$$

$$= -\frac{\dot{w}}{g}2(V_o - u) \qquad \text{[compare this to equation (4.54)]}$$

$$= -\frac{62.4 \times 0.0491 \times 4}{32.2}(2)(2) = -1.56\,\text{lb}$$

and

$$F_y = 0$$

The resultant force is 1.56 lb acting to the left.

The power transferred from a stream to a moving vane is the product of $F_x(u)$. If the turbomachine is frictionless, its output is this product, namely,

$$P = F_x(u) = \frac{\dot{w}}{g}(V_o - u)(1 - \cos\theta)u \qquad (4.56)$$

Examination of equation (4.56) shows that no power output will be obtained if the blade is either stationary or operating at $u = V_o$. If equation (4.56) is plotted as a function of u, it is found that P is maximum when $u = V_o/2$. If u is kept constant and P is plotted as a function of θ, P is found to have a maximum at $\theta = 180$ deg. Using both conditions we find that the maximum power transferred, P, is exactly equal to the power of the free jet. A blade operated at half the stream speed with the stream turned through 180 deg will have an efficiency of energy transfer theoretically equal to 100 percent.

4.8 Closure

The equations developed in Chapter 3 are the necessary relations needed to solve problems involving the steady flow of an incompressible fluid. They are widely used for this purpose and will probably be retained by the student longer than most concepts he will learn in this subject. This chapter has been devoted to the solution of many problems that can be treated analytically. Due to nonideal flow situations, empirical coefficients have been introduced to yield a better correlation between measured and computed quantities. In each application studied in this chapter the actual physical problem was reduced to an idealized one-dimensional model in which it was unnecessary to know the details of the flow in order to obtain the desired information. It will be noted that the judicious selection of the control volume makes this approach feasible and considerably reduces the required computational effort.

REFERENCES

1. *Elements of Practical Aerodynamics* by B. Jones, 4th ed., John Wiley & Sons, Inc., New York, 1950.

2. *Basic Fluid Mechanics* by J. L. Robinson, McGraw-Hill Book Company, Inc., New York, 1963.

3. *Engineering Applications of Fluid Mechanics* by J. C. Hunsaker and B. G. Rightmire, McGraw-Hill Book Company, Inc., New York, 1947.

4. *Elementary Theoretical Fluid Mechanics* by K. Brenkert, Jr., John Wiley & Sons, Inc., New York, 1960.

5. *Fluid Mechanics* by V. L. Streeter, 3rd ed., McGraw-Hill Book Company, Inc., New York, 1962.

6. *Elementary Fluid Mechanics* by J. K. Vennard, 3rd ed., John Wiley & Sons, Inc., New York, 1954.

7. *Mechanics of Fluids* by G. Murphy, 2nd ed., International Textbook Company, Scranton, Pa., 1952.

8. *Fluid Mechanics for Engineers* by P. S. Barna, Butterworth & Co. (Publishers) Ltd., London, 1957.

9. *Fluid Meters: Their Theory and Application,* 5th ed., American Society of Mechanical Engineers, New York, 1959.

10. "Flow of Fluids Through Valves, Fittings, and Pipe," *Technical Paper #410,* Crane Co., Chicago, 1957.

PROBLEMS

4.1 A jet of water issues vertically from a nozzle with a velocity of 25 ft/sec. What is the maximum height that it will reach if air resistance is neglected?

4.2 A 4-in.-i.d. pipe carries water at the rate of 300 gal/min. If the pressure in the pipe is 25 psig, determine the velocity in the pipe, the velocity head, the pressure head, and the total head referenced to datum 10 ft below the center of the pipe.

4.3 Oil (specific weight, 50 lb/ft^3) flows in a 4-in.-i.d. pipe. If the pressure in the pipe is 30 psia and the total head with respect to a reference plane 5 ft below the center of the pipe is 200 ft (of oil), determine the velocity in the pipe and the volume rate of flow in gallons per minute.

4.4 A stream flows over a waterfall 150 ft high. What is the velocity with which it strikes the base of the fall if air resistance is negligible?

4.5 In problem 4.4 assume the flow rate to be 500 ft^3/sec and determine the horsepower of the fall.

4.6 A tank has a water level of 20 ft above a datum plane. Five feet above the datum plane there exists a 6-in. opening in the tank. If there are no flow losses, determine the velocity of the water leaving the opening and the quantity of water flowing at this instant in gallons per minute.

4.7 Solve problem 4.6 if C_V is 0.95 and $C_C = 0.61$.

4.8 An airplane travels at 300 mph. If atmospheric pressure is 14.6 psia and the specific weight of air is 0.075 ft³/lb, determine the stagnation pressure exerted on the airplane.

4.9 If a propeller on an airplane going at 200 mph causes the air to leave with an absolute velocity of 280 mph, determine its efficiency.

4.10 A propeller 5 ft in diameter is mounted on a plane traveling at 350 mph. If the airstream leaves the propeller with an absolute velocity of 410 mph, determine the efficiency of the propeller and the thrust imparted to the aircraft. Assume $\gamma = 0.075/ft^3$.

4.11 If a propeller has an efficiency of 90 percent, determine its thrust if it is 6 ft in diameter and is mounted on a plane traveling at 275 mph.

4.12 Determine the horsepower absorbed by a propeller if it is on a plane traveling at 240 mph, is 7 ft in diameter, and is 80 percent efficient.

4.13 A propeller 7 ft in diameter is capable of exerting a 1500-lb thrust on an airplane traveling at 280 mph. Determine its efficiency.

4.14 A siphon is used to discharge water from one tank to another. If the level in the second tank is 10 ft below the level of the first tank, what is the maximum flow rate that can be obtained through a 3-in.-i.d. tube?

4.15 A Pitot tube is used to determine the velocity of a stream of water. If the specific weight of the water is 60 lb/ft³, and the water rise in a vertical tube attached to the Pitot tube is 5 in., determine the velocity of the stream at this point.

4.16 Derive an equation for the trajectory of a jet issuing from an orifice with a head of h ft and a velocity coefficient of unity. Assume resistance to be negligible.

4.17 A body is immersed in a stream of water in such a manner that the water is caused to stagnate against the front face of the body. If the body is 10 ft below the surface of the stream and a Pitot tube in the face of the body has a liquid rise above the stream surface of 1 ft, calculate the stream velocity at this point.

4.18 A Pitot-static tube is directed into a stream of water flowing with a velocity of 10 ft/sec. If the manometer fluid is mercury ($\gamma = 850$), determine its coefficient when the differential of the manometer is 1.8 in.

4.19 If the pressure on a submerged research vessel at a level of 500 ft (assume $\gamma = 62.4 = $ constant) is 220 psig, determine its velocity.

4.20 An airplane is designed to have a cruising speed of 500 mph. If the specific weight of air is 0.075 lb/ft³, determine the pressure between the openings of a Pitot-static tube.

4.21 A Pitot-static tube is used to measure the flow of air in a duct. If the density of the air is 0.075 lb/ft³ and a differential manometer reads 1 in. H₂O (γ = 62.4), determine the air velocity.

4.22 An airplane travels at 400 mph in air whose density is 0.07 lb/ft³. A Pitot-static tube is installed with the static connection on a wing and the Pitot connection facing forward into the air stream. If the air velocity relative to the wing is 480 mph, determine the reading of a differential pressure gage to which this instrument is connected.

4.23 A venturi meter consists of a 4-in.-diameter inlet and a 2-in.-diameter throat. If 1 ft³/sec of water is flowing in the meter, determine h in Figure 4.19 if γ is 850 lb/ft³ for mercury in the manometer.

4.24 Solve problem 4.23 if the discharge coefficient of this meter of 0.9.

4.25 A venturi meter is installed in a pipe so that its axis is vertical. An identical venturi meter is installed in a pipe the same size with its axis horizontal. If each meter gives the same reading on a differential manometer, is the flow the same in both pipes, assuming the density of the fluids to be the same for each installation?

4.26 Show that independent of the angle of the axis of a venturi meter, equation (4.47) or (4.48) holds.

4.27 Can the discharge coefficient of a venturi meter exceed unity? Assume there is friction present in the meter.

4.28 A stream of air (γ = 0.076 lb/ft³) travels horizontally with a velocity of 60 ft/sec. What force is required to deflect the air stream 180 deg if it has a cross-sectional area of 20 ft²?

4.29 A jet of water, 3 in. in diameter, strikes a flat plate placed normal to the direction of flow. If the plate is stationary, calculate the force on it if the initial velocity of the jet is 40 ft/sec.

4.30 A moving vane turns a 1-in.-diameter jet through an angle of 180 deg. If the initial absolute velocity of the jet is 100 ft/sec and the vane velocity is 70 ft/sec, determine the force on the blade.

4.31 Determine power transferred to the turbomachine of problem 4.30.

4.32 What is the maximum power that could have been transferred to the turbomachine of problem 4.30 if it were operated at optimum conditions.

NOMENCLATURE

A = area
C = empirical constant
C_C = coefficient of contraction
C_D = coefficient of discharge
C_V = coefficient of velocity
D = discrepancy
d = diameter
E = energy loss
F = force
g = local acceleration of gravity
h = head
P = power
p = pressure
Q = volume rate of flow
R = resultant force
Re = Reynolds number
T = thrust
u = velocity
V = velocity
$\dot{w}$ = weight rate of flow

x = horizontal displacement
x = height above siphon
y = height
Z = height
γ = specific weight
Δ = small change or increment
η = efficiency
θ = angle
ρ = density

Subscripts

a = atmosphere
a, b, c = locations
M = manometer
o = pipe
p = pump
T = turbine
vc = vena contracta
w = water
$1, 2, 3$ = specific states or locations

Chapter 5

DIMENSIONAL
ANALYSIS

5.1 Introduction

The physical sciences are concerned with the discovery and formulation of exact relations between the various quantities involved in recurring situations. Dimensional analysis is based upon the axiom that a general relationship may be established between two quantities only when those two quantities have the equivalent physical nature and are measured in the same units. Based upon this axiom, an equation must at least be dimensionally homogeneous (have the same dimensions on both sides of the equality) if it is to be generally valid. However, dimensional homogeneity does not guarantee that an equation is valid, nor does an equation with differing dimensions on both sides of the equality have necessarily to be invalid. Many empirical equations of this type (non-homogeneous) exist either for convenience or to express experimental data over a limited range of the variables involved. For example, the heat transfer to a vertical plate by natural convection from air is given as $q = 0.3(\Delta T)^{5/4}$, where

q is the quantity of heat transferred per unit time per unit area, British thermal units per hour-square foot; 0.3 is a dimensional constant; and ΔT is the temperature difference, degrees Fahrenheit. Note that it is correct only for the units given. This equation is nonhomogeneous, and it is therefore not general, nor is it applicable to all gases for all temperatures and pressures. However, this equation is widely used to calculate the heat transfer by natural convection from air within specified limits of temperature and pressure. The justification for using it is based upon convenience and the fact that it provides sufficient accuracy in the range of pressure and temperature to which it is limited.

To identify an observed phenomenon, an event, or a fact, it is necessary to describe the occurrence sufficiently to distinguish it from any other phenomenon, or event, or fact. The record of the observation is made in terms of certain characteristics that distinguish it and identify the class of the quantity, such as, length, force, time, etc. For the purpose of applying the principle of dimensional homogeneity, these characteristics will be defined as the dimensions of the quantity observed. It should be noted that the magnitude of a dimension serves to express its size relative to some arbitrary standard, and the dimension should not be confused with its magnitude.

In the following paragraphs of this chapter the principle of dimensional homogeneity will be applied to various physical situations. The formal procedure of applying this principle is known as dimensional analysis. The utility and importance of dimensional analysis in engineering can be seen from the following listing of some of its applications:

1. Systemizing the collection of data in an experimental program and reducing the number of variables that must be investigated.
2. Developing equations.
3. Establishing correlations based on model testing. The design of the model and its operating conditions are based upon this principle.
4. Classifying equations and indicating their generality.
5. Converting data from one system of units to another.

5.2 Systems of Units

Unfortunately there has arisen a great deal of confusion among students regarding the fundamental dimensions of force and mass. To clarify this situation, let us consider a system in which the dimension of force (F) is taken to be the fundamental unit. To relate force and mass (M) we shall use Newton's second law, namely, $F = Ma$. If acceleration (a) is expressed as the time rate of change of velocity and velocity is the time rate of change of distance, then we can relate force (F), mass (M), time (T), and length (L), by equating the dimensions on both sides of Newton's second law in terms of these quantities. Performing this operation we find that mass can be written dimensionally as

$$\frac{\text{Force} \times \text{Time}^2}{\text{Distance}}$$

Table 5.1

Symbol	Quantity	F-L-T-θ	M-L-T-θ
α	Acceleration (angular)	T^{-2}	T^{-2}
a, g	Acceleration (linear)	LT^{-2}	LT^{-2}
A	Area	L^2	L^2
ρ	Density	FT^2L^{-4}	ML^{-3}
E	Energy	FL	ML^2T^{-2}
F	Force	F	MLT^{-2}
H	Heat	H	H
L	Length	L	L
M	Mass	M	M
β	Bulk modulus	FL^{-2}	$ML^{-1}T^{-2}$
E	Modulus of elasticity	FL^{-2}	$ML^{-1}T^{-2}$
J	Mechanical equivalent of heat	FLH^{-1}	$ML^2T^{-2}H^{-1}$
M	Momentum	FT	MLT^{-1}
P	Power	FLT^{-1}	ML^2T^{-3}
p	Pressure	FL^{-2}	$ML^{-1}T^{-2}$
c	Specific heat	$HLF^{-1}T^{-2}\theta^{-1}$	$HM^{-1}\theta^{-1}$
γ	Specific weight	FL^{-3}	$ML^{-2}T^{-2}$
v	Specific volume	$L^4F^{-1}T^{-2}$	L^3M^{-1}
σ	Surface tension	FL^{-1}	MT^{-2}
T	Time	T	T
k	Thermal conductivity	$HL^{-1}T^{-1}\theta^{-1}$	$HL^{-1}T^{-1}\theta^{-1}$
t	Torque	FL	ML^2T^{-2}
α	Thermal diffusivity	L^2T^{-1}	L^2T^{-1}
β	Coefficient of thermal expansion	θ^{-1}	θ^{-1}
R_t	Thermal resistivity	$LT\theta H^{-1}$	$LT\theta H^{-1}$
θ	Temperature	θ	θ
ω	Angular velocity	T^{-1}	T^{-1}
V	Linear velocity	LT^{-1}	LT^{-1}
μ	Absolute viscosity	FTL^{-2}	$ML^{-1}T^{-1}$
γ	Kinematic viscosity	L^2T^{-1}	L^2T^{-1}
V	Volume	L^3	L^3
Q or $\dot{Q}$	Volume rate of flow	L^3T^{-1}	L^3T^{-1}
W	Weight	F	MLT^{-2}
G	Weight rate of flow	FT^{-1}	MLT^{-3}
G	Mass rate of flow	$FL^{-1}T$	MT^{-1}
N	revs per second	T^{-1}	T^{-1}

or symbolically as FT^2/L. Thus in this system, density, which is mass per unit volume, becomes

$$\frac{FT^2/L}{L^3} = FT^2/L^4 = FT^2L^{-4}$$

Since weight is simply a force, we can express weight (W) as being dimensionally equal to F. Specific weight, the weight per unit volume, is therefore F/L^3. Using force as a fundamental dimension, an entire system of consistent dimensions has been devised which is known as the F-L-T-θ system, where θ denotes temperature. Some important physical quantities are listed in this system in Table 5.1.

The choice of force as a fundamental dimension is arbitrary, and instead we could have chosen mass (M) as a fundamental dimension. If mass had been selected as a fundamental dimension, we would relate it to the other dimensions by again invoking $F = Ma$. In this system the fundamental dimensions are mass, length, time, and temperature, and the derived dimension of force has the units of MLT^{-2}. Table 5.1 also shows the M-L-T-θ system of dimensions. In both of the systems shown in Table 5.1 the independent symbol H is used for the quantity heat. Since heat is a form of energy, the symbol H is not an independent quantity and can be replaced by FL or ML^2T^{-2} in the F-L-T-θ or M-L-T-θ systems, respectively. Other systems can be chosen, but these two systems are most commonly used since they involve physical quantities for which standards of measurement exist and are convenient for comparison.

5.3 Pi Theorem

Let us consider a phenomenon where one of the variables (the dependent variable) is controlled by several other variables. A change in these variables (independent variables) will influence the dependent variable but will not influence any of the independent variables. Mathematically we can express this requirement as

$$F = f(E_1, E_2, E_3, \ldots, E_N) \tag{5.1}$$

where F is the dependent variable; $E_1, E_2, E_3, \ldots, E_N$ are the independent variables; and f symbolically means "a function of." A relation such as equation (5.1) can usually be rewritten as a series of a number of terms each composed of the product of the independent variables raised to a suitable power. Thus

$$F = E_1^{a_1} E_2^{b_1} E_3^{c_1} \cdots E_N^{r_1} + E_1^{a_2} E_2^{b_2} E_3^{c_2} \cdots E_N^{r_2} + \cdots \qquad (5.2)$$

where the exponents a, b, and c are numbers only (without dimensions). If we now divide both sides of equation (5.2) by $E_1^{a_1} E_2^{b_1} E_3^{c_1} \cdots E_N^{r_1}$, we obtain,

$$\frac{E}{E_1^{a_1} E_2^{b_1} E_3^{c_1} \cdots E_N^{r}} = 1 + E_1^{(a_2-a_1)} E_2^{(b_2-b_1)} E_3^{(c_2-c_1)} \cdots E_N^{(r_2-r_1)} \qquad (5.3)$$

Since the first term on the right-hand side is unity (dimensionless), dimensional homogeneity requires that all of the terms in equation (5.3) must be dimensionless. The left-hand side of the equation is a dimensionless group that we shall now denote as π_1. Similarly, we may form other dimensionless groups π_2, $\pi_3, \pi_4, \ldots, \pi_{N-M}$ by dividing equation (5.2) successively by each of the groups on the right-hand side. The resulting equation describing the phenomenon will now have the form,

$$f'(\pi_1, \pi_2, \pi_3, \ldots, \pi_{N-M}) = 0$$

where f' is used to denote a new function of the indicated variables. If the number of physical dimensions involved in the N quantities or entities is J, the minimum number of dimensionless groups required to describe the phenomenon is,

$$M = N - J \qquad (5.4)$$

Table 5.2 Dimensionless Groups

Symbol	Name	Group†
Bi	Biot Number	$hD/2K$
Fo	Fourier number	$KT/\rho c D^2$
Fr	Froude number	$V/\sqrt{Lg}$ or V^2/Lg
Gr	Grashof number	$D^3 \rho^2 g \beta (\Delta\theta)/\mu^2$
Nu	Nusselt number	hD/k
Pe	Peclet number	$DV\rho c/K$
Pr	Prandtl number	$c\mu/k$
Re	Reynolds number	$DV\rho/\mu$
Sc	Schmidt number	$\mu/\rho k$
St	Stanton number	$h/cV\rho$
We	Weber number	$\rho L V^2/\sigma$
C	Cauchy number	$\rho V^2/\beta$
M	Mach number	$V/\sqrt{\beta/\rho}$ or $M = \sqrt{C}$

†V is velocity, θ is a temperature difference, and D is diameter. Other symbols are given in Table 5.1. In this table β is the coefficient of thermal expansion.

This relation is known as the *pi theorem*, and its application enables us to reduce the minimum number of independent variables from N to $N - J$. The basic restriction placed on the pi terms is that they must be dimensionless and independent. For example, say there is an event in which it is known that five variables expressible in terms of three fundamental dimensions are involved. In this case $N - J = 5 - 3 = 2$, and a relationship can be developed in terms of two pi terms.

Certain groupings of terms have been named in honor of various investigators and some of these groupings are listed in Table 5.2. The utility of some of these groups will become evident later in this chapter.

5.4 Dimensional Analysis

The application of the principle of dimensional homogeneity complemented by the statement of the pi theorem constitutes the subject of dimensional analysis. However, it is important to note that even though dimensional analysis is a powerful and useful tool it has certain restrictions:

1. No insight into the mechanism of the process can be gained directly from it.
2. No clue is provided as to whether the quantities supposed to be involved are sufficient or even significant; if they are not, the resultant form deduced for the dimensionless groups may not correspond with the results deduced from experiments.
3. No clue is provided as to which groups are dominant and which ones are not significant.
4. The values of the dimensionless groups at which critical or abrupt changes in the character of the processes take place are not predicted by dimensional analysis.
5. The numerical value of exponents is unknown and must be determined either from experiments or from the differential equation governing the process.

The student will find that the practical application of the principle of dimensional analysis to problems can be considerably simplified by adapting the following procedure:

1. List all independent variables that experience shows enter into the problem. This step is the most important step in the procedure, and it is the most difficult. Any test results, preliminary mathematical analyses, or even simply physical intuition should be invoked at this point.
2. Assign symbols to the independent variables selected in step 1. It is usually convenient to use the same symbols as in Table 5.1.
3. Write a functional equation for the phenomenon in terms of the symbols as-

signed in step 2. Assign arbitrary exponents to each of the independent variables.
4. Substitute the appropriate dimensions for each of the variables in step 3. Retain the exponents in the equation and note that the exponent of the dependent variable is unity.
5. Write simultaneous algebraic equations equating exponents of like dimensions.
6. Solve the equations of step 5 in terms of the exponents.
7. Express the general equation in terms of the results of step 6 and collect and group the variables into dimensionless groups.
8. The results of steps 7 should be checked against the pi theorem.

ILLUSTRATIVE PROBLEM 5.1. To illustrate the method of dimensional analysis, consider the motion of a freely falling body. We would like to develop the form of the equation expressing the distance traveled by the body in a given time, T.

Solution. Step 1. For this problem it will be assumed that the distance traveled by the body is known to depend on the time the body has fallen and the acceleration of gravity. In words, we write the functional relation

$$\text{Distance} = f\{(\text{Acceleration of gravity})^a \, (\text{Time})^b\}$$

Step 2. Use the symbols given in Table 5.1.

Step 3. Symbolically,

$$s = \text{Constant } \{(g)^a (T)^b\}$$

Step 4. Dimensionally (in either the M-L-T-θ or F-L-T-θ systems),

$$(L) = (LT^{-2})^a (T)^b$$

Step 5. Equating exponents,

1. Mass, $0 = 0$.
2. Length, $1 = a$.
3. Time, $0 = -2a + b$.

Step 6. Solving these equations we obtain $a = 1$ and $b = 2$.

Step 7. The derived relation is therefore $s = \text{constant } (gT^2)$, where the constant cannot be determined by dimensional analysis since it is dimensionless. At this point either experiments must be performed or the appropriate differential equation must be solved to evaluate the dimensionless constant.

Step 8. This step is left as an exercise for the student.

Let us now assume that the distance traveled by the freely falling body is also a function of the mass of the body. The word equation becomes,

Distance = Constant $\{(\text{Acceleration of gravity})^a\ (\text{Time})^b\ (\text{Mass})^c\}$

or dimensionally, in the $M\text{-}L\text{-}T\text{-}\theta$ system,

$$(L)^1 = (LT^{-2})^a (T)^b (M)^c$$

Equating exponents,

1. Mass, $0 = c$.
2. Length, $1 = a$.
3. Time, $0 = -2a + b$.

Therefore $a = 1$, $b = 2$, and $c = 0$.

The derived relation for this case is $s = \text{constant}\ (gT^2)$. In this case the additional variable, mass, is not independent, and the solution is found to be the same as when mass was omitted from the analysis. However, it is not always true that the introduction (or deletion) of a variable will lead to a meaningful result. This point is of extreme importance and cannot be overemphasized. In any problem it is important to include all relevant variables and to exclude all unnecessary ones. If too few quantities are included, the solution will be incomplete and probably useless; if too many quantities are included, there is no saving in overall effort. Unfortunately the setting up of a problem in dimensional analysis requires an intuitive insight into the physical problem and intuition can often be misleading.

ILLUSTRATIVE PROBLEM 5.2. As another illustration of the application of dimensional analysis let us consider the following physical situation. A weight W is hung from a spring that has a spring constant (restoring force per unit displacement) k. If the spring is initially given a displacement x_o and subsequently released and allowed to oscillate free, it is desired to obtain an expression for the frequency (inverse of the period) of the vibrating system. A schematic sketch of the system is shown in Figure 5.1. We shall initially

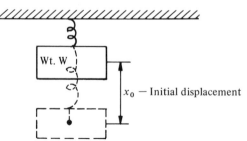

Wt. W

x_0 — Initial displacement

Figure 5.1. Simple vibrating system.

assume that the restoring force per unit displacement (k) is a constant and describe this as being a linear spring.

Solution. The first step in solving this problem is to decide on the independent variables that enter the problem. In this case we shall take the weight of the body, the spring constant, and the acceleration of gravity as being the items of importance. Since the spring was assumed to be linear, we shall assume that the initial displacement x_o does not enter the problem. This point will be considered later on. Applying the pi theorem [equation 5.4] we have

$$M = N - J = 4 - 3 = 1 \qquad \text{(frequency is included as a variable in } N)$$

and we can expect that a minimum of one dimensionless group will serve as a solution to the problem.

Setting up a power series and using words for each of the physical quantities,

$$\text{Frequency} = \frac{1}{\text{Period}} = f$$

$$= \text{Constant (Weight)}^a \text{ (Spring constant)}^b \text{ (Acceleration of gravity)}^c$$

and

$$f = \text{Constant } (W)^a (k)^b (g)^c$$

Using the M-L-T-θ system and Table 5.1,

$$T^{-1} = (MLT^{-2})^a \left(\frac{MLT^{-2}}{L}\right)^b (LT^{-2})^c$$

At this point we again note that dimensional homogeneity requires that the dimensions on both sides of this equation must be equal. Equating exponents,

1. Mass, $a + b = 0$.
2. Time, $-2a - 2b - 2c = -1$.
3. Length, $a + c = 0$.

Solving these equations simultaneously, we obtain

$$a = -c, \qquad b = c, \qquad c = \tfrac{1}{2}$$

Returning to the original word equation,

$$\frac{1}{\text{Period}} = \text{(Weight)}^{-1/2} \text{ (Spring constant)}^{1/2} \text{ (Acceleration of gravity)}^{1/2}$$

Collecting terms,

$$\text{Frequency } (f) = \text{Constant } \left(\sqrt{\frac{kg}{W}} \right)$$

Thus a spring with a constant value of k will have a frequency proportional to $\sqrt{kg/W}$. Note that we are unable to determine the constant of proportionality since this can be found only by experiment or from the solution of the appropriate differential equation.

Let us now assume that the initial displacement (x_o) is an additional independent variable. From the pi theorem we have the requirement that at least two dimensional groups will be necessary for the solution of this problem since $M = N - J = 5 - 3 = 2$. Proceeding as before,

$$\frac{1}{\text{Period}} = \text{Constant (Weight)}^a \text{ (Spring constant)}^b \text{ (Acceleration of gravity)}^c$$

$$(\text{Displacement})^d$$

or dimensionally,

$$T^{-1} = (MLT^{-2})^a \left(\frac{MLT^{-2}}{L} \right)^b (LT^{-2})^c (L)^d$$

1. Mass, $a + b = 0$.
2. Length, $a + c + d = 0$.
3. Time, $-2a -2b -2c = -1$.

Solving,

$$c = \tfrac{1}{2}$$
$$a = -d - \tfrac{1}{2}$$
$$b = d + \tfrac{1}{2}$$

Consequently,

$$T^{-1} = (\text{Weight})^{-d-1/2} \text{ (Spring constant)}^{d+1/2} \text{ (Acceleration of gravity)}^{1/2}$$

$$(\text{Displacement})^d$$

or

$$\text{Frequency} = \text{Constant } \sqrt{\frac{kg}{W}} \left(\frac{k}{W} x_o \right)^d$$

At this point we cannot determine from dimensional analysis whether the term $(kx_o/W)^d$ is or is not important to the final solution. Again it is necessary to resort to either experiments or the appropriate differential equation. It is interesting to note that for a linear spring, the frequency is not a function of the initial displacement and the exponent d in the equation becomes zero. However, if the spring is nonlinear (k is not constant), it has been found that the term $(kx_o W)^d$ must be retained in the final solution.

ILLUSTRATIVE PROBLEM 5.3. An incompressible fluid flows in a long pipe. Derive a relationship between the variables involved in this phenomenon to express the pressure loss in the pipe due to fluid flow.

Solution. As was noted earlier, the selection of the quantities that enter a problem represents the most difficult part of the dimensional analysis procedure. For this problem, K. Brenkert, Jr. (reference **2**) uses a form of reasoning that illustrates how one selects the independent variables in a general problem. The quantities chosen and his reasoning behind the selection of each quantity is given below:

Quantity	Reasoning
Pressure loss, p	What you are solving for
Diameter of pipe, D	Certainly a variable and it could affect pressure drop
Length of pipe, L	The longer the pipe, the greater the losses
Velocity of fluid, V	Friction depends on relative velocity
Viscocity of liquid, μ	This could seem necessary in any problem involving friction
Gravitational acceleration, g	To take care of elevation changes in the pipe
Density of fluid, ρ	In problems of motion, add the mass for inertia forces
Roughness of pipe, e	Rougher pipes should cause greater losses

It is interesting to note that Brenkert points out "that the reasoning seems vague in some cases, the typically weakest point of the dimensional analysis approach."

Continuing with the problem, we may therefore write

Pressure loss $= f\{$(Pipe diameter)^a (Viscosity)^b (Velocity)^c (Density)^d

(Roughness)^e (Gravitational acceleration)^f (Length)$^g\}$

Mathematically, we have

$$p = f[(D)^a(\mu)^b(V)^c(\rho)^d(e)^e(g)^f(l)^g]$$

In M-L-T-θ units,

$$ML^{-1}T^{-2} = (L)^a(ML^{-1}T^{-1})^b(LT^{-1})^c(ML^{-3})^d(L)^e(LT^{-2})^f(L)^g$$

Equating dimensions,

1. Mass, $1 = b + d$.
2. Length, $-1 = a - b + c - 3d + e + f + g$.
3. Time, $-2 = -b - c - 2f$.

Solving these equations simultaneously in terms of b, we have

$$a = -b + f - e - g$$
$$c = 2 - b - 2f$$
$$d = 1 - b$$

Therefore,

$$p = \text{Constant } (D^{-b+f-e-g})(\mu)^b(V^{2-b-2f})(\rho^{1-b})(e^e)(g^f)(L^g)$$

and regrouping terms,

$$\frac{p}{\rho V^2} = \text{Constant } \left(\frac{DV\rho}{\mu}\right)^{-b}\left(\frac{V^2}{gD}\right)^{-f}\left(\frac{e}{D}\right)^e\left(\frac{L}{D}\right)^g$$

It is interesting to note that $\frac{1}{2}\rho V^2$ is a dynamic pressure due to bringing the flow to rest.

The dimensionless terms $DV\rho/\mu$ and V^2/gD that appear in the solution of this illustrative problem have special significance and have been named the Reynolds number and Froude number, respectively. Both of these dimensionless terms can be interpreted physically, and when this is done it is found that the Reynolds number is the ratio of the inertia to the friction force and that the Froude number is the ratio of the inertia force to the force of gravity. The Froude number is found to be the significant parameter of correlation if there is a free surface affected by the force of gravity, as in the form of waves.

ILLUSTRATIVE PROBLEM 5.4. A body is fully submerged in a viscous incompressible fluid and is moving with a velocity V. Formulate a relation between the variables involved in this process to express the drag force on the body.

Solution. Based upon any knowledge we may have of this motion, our best judgment, and intuition, we shall assume that the drag force on the body is a function only of the velocity of the body, a linear dimension (in the case of a sphere, its diameter), the density of the fluid, and the viscosity of the fluid. The word equation is therefore

Drag force = Constant (Length)^a (Velocity)^b (Density)^c (Viscosity)^d

and

$$F = \text{Constant } (D)^a(V)^b(\rho)^c(\mu)^d$$

Dimensionally in the M-L-T-θ system,

$$MLT^{-2} = (L)^a(LT^{-1})^b(ML^{-3})^c(ML^{-1}T^{-1})^d$$

Applying the requirement that the resulting equation must be dimensionally homogeneous and equating exponents,

1. Mass, $1 = c + d$.
2. Length, $1 = a + b - 3c - d$.
3. Time, $-2 = -b - d$.

Solving these equations in terms of d,

$$b = 2 - d$$
$$c = 1 - d$$
$$a = 2 - d$$

Therefore,

$$F = \text{Constant } (D)^{2-d}(V)^{2-d}(\rho)^{1-d}(\mu)^d$$

Rearranging and grouping terms,

$$F = \text{Constant } \left(\frac{DV\rho}{\mu}\right)^{-d}(\rho D^2 V^2)$$

or

$$F/\rho D^2 V^2 = \text{Constant } \left(\frac{DV\rho}{\mu}\right)^{-d}$$

Applying the pi theorem $M = N - J = 5 - 3 - 2$. Thus a minimum of two dimensionless groups has provided a solution to the problem, and we have reduced the number of unknowns from five to two.

5.5 Similarity and Model Testing

Models are often tested in wind tunnels and towing basins in order to determine the performance of full-sized aircraft and ships. Model testing has been resorted to in most cases because the fluid flow is so complicated that it is impossible to obtain an analytic solution to the problem. It therefore becomes necessary to combine a simplified mathematical analysis with experimental data in order to adequately describe some of these phenomena. Based upon tests of scale models, data can be obtained from which the performance of full-scale equipment can be predicted. The application of dimensional analysis to a problem permits us to use the derived dimensionless groups in a systematic manner to predict performance on any other scale.

To obtain quantitative results from model studies that will be applicable to the full-sized body (prototype), it is necessary that there be similarity between the model and the prototype. Similarity will be used to mean that the shape, the motion, and the forces must each have a constant (but not necessarily the same) ratio between the model and the prototype. We shall denote these three forms of similarity as (1) geometric similarity, (2) kinetic similarity, and (3) dynamic similarity.

1. For geometric similarity to exist it is necessary for the model to be an exactly scaled replica of the prototype. This should include surface roughness, rivets, etc. Thus it would be expected that the surface finish on the model would be smoother than on the prototype. To illustrate further this requirement let us return to illustrative problem 5.3 where we considered the flow of an incompressible fluid in a long pipe. By applying dimensional analysis to this problem we arrived at a solution that contained two dimensionless groups involving only the dimensions of length; these groups were the length to diameter ratio L/D and the relative roughness e/D. If geometric similarity exists, then

$$\left(\frac{L}{D}\right)_m = \left(\frac{L}{D}\right)_p \quad \text{and} \quad \left(\frac{e}{D}\right)_m = \left(\frac{e}{D}\right)_p$$

where the subscript m denotes model and the subscript p denotes prototype. These relations reduce to

$$\frac{L_m}{L_p} = \frac{D_m}{D_p} = \frac{e_m}{e_p}$$

and in general it can be concluded that the ratio of all corresponding lengths must be the same for the model and prototype to meet the requirements for geometric similarity.

2. To illustrate kinematic similarity, consider the flow around a moving body completely immersed in a viscous incompressible fluid. The condition for kinematic similarity is that the motion of the model and prototype (as well as the fluid particle motions) must be similar. All components of velocity and acceleration of the model must correspond to those of the prototype. Defining the geometric similarity factor as $k_g = L_p/L_m$, the velocity similarity factor as $k_v = V_p/V_m$, the acceleration similarity factor as $k_a = a_p/a_m$, and the time scale factor as $k_t = T_p/T_m$, we arrive at the following requirements for kinematic similarity.

$$k_g = k_v k_t; \qquad k_g = k_a k_t^2 \tag{5.5}$$

If a model is constructed to a one-tenth scale and has the same velocity as the prototype, it must have a time scale for all physical events of one-tenth that of the prototype. If the time scales are made the same, then the velocity of the model must be one-tenth that of the prototype.

3. Dynamic similarity requires that the ratio of the dynamic pressures at corresponding points of the model and prototype must be a constant. In addition, the ratios of all of the various types of forces must be the same at corresponding points of the model and prototype if the dynamic pressures are to be in the same ratio at these corresponding points.

Using the previous notation and denoting the ratio of forces as $k_f = F_p/F_m$ and the ratio of masses as $k_m = M_p/M_m$, the condition for dynamic similarity is

$$k_f = \frac{k_m k_g}{k_t^2} \tag{5.6}$$

All of the foregoing can be generalized to any flow, and paraphrasing I. H. Shames (see reference 11), "dimensional analysis will yield the dimensionless groups in a flow for which we must duplicate at least all but one in geometrically similar flows to achieve dynamic similarity." By way of illustrating this principle let us consider illustrative problem 5.3 again. The four dimensionless groups on the right-hand side of the relation derived for this problem were $DV\rho/\mu$, the Reynolds number; $V^2/(gD)$, the Froude number; e/D, the relative roughness; and L/D, the length to diameter ratio. For the flow in a model system to be similar to the flow in a prototype system we require these four

dimensionless groups to be equal in both systems. If this is done, then the remaining dependent group, $p/\rho V^2$, will also be the same in both systems. Unfortunately, strict fulfillment of the equality of these four terms is impossible
unless the model and prototype are made to the same scale and the flows are
identical. This would obviously defeat the purpose of model testing, but fortunately in most cases one effect is usually dominant over the others, and it is
possible to retain the dimensionless groups incorporating this effect and to make
corrections for the dimensionless groups that are not equal in the model and
prototype. In this problem of an incompressible fluid flowing in a long pipe,
the Reynolds number expresses the ratio of inertia force to friction force, and
for similarity the necessary and sufficient condition is that the geometric ratio
between the model and prototype be constant and that the Reynolds number
of both systems be equal. Since similarity is met, it follows that $p/\rho V^2$ must
be equal for both the model and the prototype. Therefore,

$$\left(\frac{p}{\rho V^2}\right)_m = \left(\frac{p}{\rho V^2}\right)_p$$

Since pressure is force per unit area, p can be written dimensionally as F/L^2.
Therefore,

$$k_f = \frac{F_p}{F_m} = \frac{V_p^2}{V_m^2}\frac{\rho_p}{\rho_m}\frac{L_p^2}{L_m^2} = \frac{k_m k_g}{k_t^2}$$

This meets our previously derived criteria for dynamic similarity.

ILLUSTRATIVE PROBLEM 5.5. It is desired to predict the pressure drop in a
large air duct. A model is constructed with linear dimensions one-tenth those
of the prototype, and water is used as the test fluid. If water is 1000 times denser
than air and has 100 times the viscosity of air, determine the pressure drop
in the prototype for the condition corresponding to a pressure drop of 10 psi
in the model.

Solution. For similarity the Reynolds number in the prototype and model
must be equal. Therefore,

$$\left(\frac{DV\rho}{\mu}\right)_p = \left(\frac{DV\rho}{\mu}\right)_m$$

and

$$\frac{V_p}{V_m} = \left(\frac{\rho_m}{\rho_p}\right)\left(\frac{D_m}{D_p}\right)\left(\frac{\mu_p}{\mu_m}\right) = \left(\frac{1000}{1}\right)\left(\frac{1}{10}\right)\left(\frac{1}{100}\right) = 1$$

As noted earlier, $p_p/p_m = (\rho_p/\rho_m)(V_p/V_m)^2$. Thus

$$\frac{p_p}{p_m} = \left(\frac{1}{1000}\right)\left(\frac{1}{1}\right)^2$$

and

$$p_p = 10 \text{ psi} \left(\frac{1}{1000}\right)\left(\frac{1}{1}\right)^2 = 0.01 \text{ psi}$$

ILLUSTRATIVE PROBLEM 5.6. A fully submerged sphere is dropped in water. It is desired to predict the drag force on the sphere when it reaches its terminal speed by testing a model sphere in air. If the model is made one twenty-fifth the size of the prototype and the ratio of the properties of air to water are the same as in illustrative problem 5.5, determine the drag force on the spherical prototype if it is found that the drag force on the model is 2 lb.

Solution. The dimensional analysis of this problem was discussed in illustrative problem 5.4, where the relation

$$\frac{F}{\rho D^2 V^2} = \text{Constant} \left(\frac{DV\rho}{\mu}\right)^{-d}$$

was derived. For similarity, we once again require that the Reynolds number of the prototype and the model be equal. Thus

$$\left(\frac{DV\rho}{\mu}\right)_p = \left(\frac{DV\rho}{\mu}\right)_m$$

and

$$\frac{V_p}{V_m} = \left(\frac{\rho_m}{\rho_p}\right)\left(\frac{D_m}{D_p}\right)\left(\frac{\mu_p}{\mu_m}\right) = \left(\frac{1}{1000}\right)\left(\frac{1}{25}\right) \times \left(\frac{100}{1}\right) = \frac{1}{250}$$

Thus

$$\left(\frac{F}{\rho D^2 V^2}\right)_m = \left(\frac{F}{\rho D^2 V^2}\right)_p$$

$$\frac{F_m}{F_p} = \left(\frac{\rho_m}{\rho_p}\right)\left(\frac{D_m}{D_p}\right)^2\left(\frac{V_m}{V_p}\right)^2 = \left(\frac{1}{1000}\right)\left(\frac{1}{25}\right)^2 \times (250)^2 = 0.10$$

Therefore,

$$F_p = (10)(F_m) = (10)(2 \text{ lb}) = 20 \text{ lb}$$

ILLUSTRATIVE PROBLEM 5.7. A ship model is towed in a towing basin. Based upon the results of these towing tests it is desired to determine the resistance of the prototype vessel. Using the methods of dimensional analysis, determine the requirements (in terms of dimensionless ratios) for dynamic similarity.

Solution. The physical quantities that enter the solution of this problem will be taken to be the velocity of the vessel V, the density of the fluid ρ, the viscosity of the fluid μ, a characteristic length L, the acceleration of gravity g, and the drag force F. From the pi theorem we find that a minimum of three dimensionless groups is required for the complete solution of this problem. Proceeding as before, we find

$$F = \text{Constant} \left(\frac{V^2}{Lg}\right)^a \left(\frac{LV\rho}{\mu}\right)^b \rho L^2 V^2$$

or

$$\frac{F}{\rho L^2 V^2} = \text{Constant} \left(\frac{V^2}{Lg}\right)^a \left(\frac{LV\rho}{\mu}\right)^b$$

The first term of the right-hand side is the Froude number, and the second term is the Reynolds number. If the model is towed in the same fluid as the prototype, dynamic similarity requires $(LV)_m = (LV)_p$ and $(V^2/L)_m = (V^2/L)_p$. The only way that these requirements can be met is for $L_m = L_p$, or the model must be full scale. It is possible to use other fluids, but these are not economically feasible; we must therefore conclude that complete dynamic similarity is not possible. If wave motion predominates (due to the pressure of a free surface) over skin friction, we may scale the model to make the Froude number the same for both the model and prototype. In this event, $(V^2/L)_m = (V^2/L)_p$ and $(F/\rho L^2 V^2)_m = (F/\rho L^2 V^2)_p$, and from these relations the prototype can be evaluated in terms of the model. Corrections can be made to these results to account for the skin friction effects. When some of the dimensionless terms cannot be made equal in the model and the prototype, as in this problem, we have incomplete dynamic similarity. For a further discussion of incomplete dynamic similarity, the references should be consulted.

ILLUSTRATIVE PROBLEM 5.8. A model of the hull of the Queen Elizabeth II (QEII) is to be constructed to determine the drag characteristics of the QEII in a towing tank. If the hull model is made 10 ft long and the ship is taken to be 1000 ft long, at what speed should the model be operated to obtain data for the QEII operating at 30 knots. Assume that the towing tank is filled with sea water.

Solution. From illustrative problem 5.7, we have

$$\frac{F}{\rho L^2 V^2} = \text{Constant} \left(\frac{V^2}{Lg}\right)^a \left(\frac{LV\rho}{\mu}\right)^b$$

As noted earlier in illustrative problem 5.7, when wave motion predominates over skin friction we may use the condition that the Froude number must be the same for the model and prototype. Therefore,

$$\left(\frac{V^2}{L}\right)_m = \left(\frac{V^2}{L}\right)_p$$

$$\frac{V_m^2}{V_p^2} = \frac{L_m}{L_p} = \frac{10}{1000} = \frac{1}{100}$$

and

$$V_m = \frac{1}{10}V_p = \frac{30}{10} = 3 \text{ knots}$$

The force requirement is given by

$$\left(\frac{F}{\rho L^2 V^2}\right)_m = \left(\frac{F}{\rho L^2 V^2}\right)_p$$

$$\frac{F_m}{F_p} = \left(\frac{\rho_m}{\rho_p}\right)\left(\frac{L_m}{L_p}\right)^2\left(\frac{V_m}{V_p}\right)^2 = \frac{1}{1} \times \left(\frac{10}{1000}\right)^2 \times \left(\frac{3}{30}\right)^2 = \frac{1}{10,000} \times \frac{1}{100}$$

$$= 10^{-6}$$

Therefore, $F_m = 10^{-6}F_p$, and the forces on the model will be one millionth of the forces on the actual ship. It is imperative that the hull model be made very accurately if such a force scale factor is to be meaningful.

5.6 Closure

In this chapter we have discussed the principle of dimensional homogeneity and noted that this principle enables us to evaluate the consistency or completeness of any correlation or analysis of physical events. Dimensional analysis, which is the application of the principle of dimensional homogeneity, furnishes the general condition that must exist between the variables involved and yields the form of the mathematical relation between these variables. It has been noted, however, that dimensional homogeneity does not necessarily make an equation valid; it is a necessary but not sufficient condition for the general validity of any equation.

Model testing is used extensively to evaluate the performance and design parameters of many devices. The experimentally determined data can then be used to predict the performance of other similar devices on another scale. For the model and prototype to be similar we required that the model and prototype be geometrically, kinematically, and dynamically similar. Complete similarity is usually not possible, and it is usually necessary to plan a given test carefully and to concentrate on the predominant phenomena, making corrections as required for other effects.

Dimensional analysis and model testing are both valuable tools when properly used. However, when applied incorrectly or without an understanding of the physical events involved, these tools (like all other mishandled tools) can prove to be liabilities.

REFERENCES

1. *Basic Fluid Mechanics* by J. L. Robinson, McGraw-Hill Book Company, Inc., New York, 1963.

2. *Elementary Theoretical Fluid Mechanics* by K. Brenkert, Jr., John Wiley & Sons, Inc., New York, 1960.

3. *Engineering Applications of Fluid Mechanics* by J. C. Hunsaker and B. G. Rightmire, McGraw-Hill Book Company, Inc., New York, 1947.

4. *Process Heat Transfer* by D. Q. Kern, McGraw-Hill Book Company, Inc., New York, 1950.

5. *Fluid Mechanics* by R. C. Binder, 4th ed., Prentice-Hall, Inc., Englewood Cliffs, N. J., 1962.

6. *Heat Transfer Notes* by John Blizard, revised by G. E. Tate, Polytechnic Institute of Brooklyn, Brooklyn, 1951.

7. *Similitude in Engineering* by G. Murphy, The Ronald Press Company, New York, 1950.

8. "Techniques of Dimensional Analysis" by W. T. Illingworth, *Electro-Technology (New York)*, 75, March 1965.

9. "Fluid Mechanics" by A. D. Kraus, *Electro-Technology (New York)*, 69, April 1962, p. 120.

10. *Fluid Mechanics for Engineers* by P. S. Barna, Butterworth & Co. (Publishers), Ltd., London, 1957.

11. *Mechanics of Fluids* by Irving H. Shames, McGraw-Hill Book Company, Inc., New York, 1962.

12. *Fluid Mechanics Through Worked Examples* by D. R. L. Smith and J. Houghton, Dover Publications, Inc., 1960.

13. "Dimensional Analysis" by R. A. Deutsch, *Electro-Technology (New York)*, 70, August 1962, p. 107.

PROBLEMS

5.1 Solve illustrative problem 5.1 using the F-L-T-θ system.

5.2 Solve illustrative problem 5.2 using the F-L-T-θ system.

5.3 Solve illustrative problem 5.3 using the F-L-T-θ system.

5.4 Solve illustrative problem 5.4 using the F-L-T-θ system.

5.5 A simple pendulum consists of a mass m hung on a string of length l. Using the method of dimensional analysis, evalute the period of the pendulum. Assume that for small angular displacements the restoring force on the pendulum is proportional to the angular displacement. Essentially this is the same as a spring with a linear spring constant. Include the acceleration of gravity in the analysis.

5.6 In static liquids it is known that the pressure at any depth varies directly with the depth of the fluid and its specific weight. By use of dimensional analysis, obtain the form of the equation relating these variables.

5.7 The speed of a small disturbance in gas is a function of the specific weight of the gas, the initial pressure of the gas, and the viscosity of the gas. Evaluate an expression relating the speed of the disturbance as a function of these variables and g, the acceleration of gravity.

5.8 The terminal velocity of a sphere dropped in a fluid is known to be a function of the diameter of the sphere, the specific weight of the fluid, and the viscosity of the fluid. Using dimensional analysis, relate the terminal velocity of the sphere to these variables.

5.9 A beam is loaded by a force F. Relate the deflection of a beam to F, L, E, and I, where L is the length of the beam, E the modulus of elasticity (force per unit area), and I the second moment of area (commonly called the moment of inertia, units of length^4).

5.10 A centrifugal compressor requires a certain amount of power at the wheel of the compressor. If the diameter of the wheel, the density of the fluid, the pressure at the outlet of the compressor, the speed of the compressor, and the acceleration of gravity are the variables involved, evaluate a functional relation for the pump power.

5.11 The Reynolds number in a pipe flowing full of liquid is 75,000. If the fluid is heated so that the viscosity is halved and the density is decreased by a factor of 3, what is the new Reynolds number?

5.12 It is desired to determine the resistance to flow for water in a 6-in.-diameter pipe. However, there is only a piece of 1-in.-diameter pipe available. If the water is at the same temperature in both instances, what should the ratio of the velocity in the 1-in.-diameter pipe to the 6-in.-diameter pipe be to obtain similiarity?

5.13 A submarine model is to be tested to determine the submerged characteristics of the ship. If the model is made to a scale of 1 in. = 1 ft and is towed at 1 ft/sec, what is the corresponding speed of the prototype?

5.14 In illustrative problem 5.4, N is taken to be 5. State explicitly the five items that make up N for this problem.

5.15 If the submarine in problem 5.13 is operated as a surface ship, determine the corresponding prototype speed.

5.16 The specific weight of air at room temperature and pressure (60°F and 14.7 psia) is 0.075 lb/ft³. The viscosity of air can be taken as a function of temperature only. If the pressure in a wind tunnel is 10 times atmospheric pressure and the temperature is 60°F, the specific weight of air will be 0.375 lb/ft³. Determine the model speed if a model is to be made to a scale of 1 in. = 1 ft and is to simulate a prototype airplane operating at 350 mph.

5.17 Air flows through a 6-in.-diameter pipe at an average velocity of 150 ft/sec. If the flow in an 8-in.-diameter pipe is to be dynamically similar to the flow in the 6-in.-diameter pipe, compute the average velocity in the 8-in. pipe. Assume the air to be at the same pressure and temperature in both cases.

Chapter 6

STEADY FLOW OF INCOMPRESSIBLE FLUIDS IN PIPES

6.1 Introduction

Fluids are most commonly transported from one location to another by forcing them through pipes and tubes. The Bernoulli equation that was developed in Chapter 3 and extensively applied in Chapter 4 is applicable to the flow of a frictionless fluid and with modification can be applied (judiciously) to the flow of real fluids by inserting a term in the equation to account for these real effects (losses). In this chapter we shall consider the problem of the steady flow of incompressible fluids in pipes, define the flow regimes encountered and the general form of the friction equation, and give expressions for entrance, exit, enlargement, contraction, bend, and valve losses. The literature on this subject is quite extensive, and the data relating to these effects have been presented in many different (and many times inconsistent) forms. One of the most comprehensive, consistent correlations of available data has been carried out by the Hydraulic Institute and has been published as the Pipe Friction section of the *Standards of the Hydraulic Institute.* Several of the tables and charts in this

161

chapter are taken from this standard with permission of the Hydraulic Institute, and for more complete and extensive literature references and details the student is specifically referred to this publication in addition to the other general references given at the end of this chapter.

6.2 Character of Flow in Pipes—Laminar and Turbulent

To visualize the character of a fluid flowing in a pipe, Osborne Reynolds devised a simple experiment, shown schematically in Figure 6.1. Basically the dye is introduced into the glass tube by injecting it from a fine piece of tubing

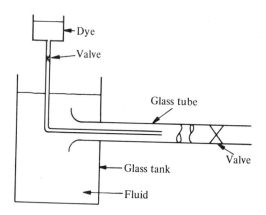

Figure 6-1. Schematic of Reynolds' apparatus.

into the well-rounded entrance of the glass tube. The velocity of the test fluid is controlled by varying the height of the fluid in the glass tank and by changing the setting of the valve in the downstream portion of the glass tube.

At small average velocities it is found that the dye filament appears as a straight unbroken line parallel to the axis of the tube. This type of flow is known as laminar, viscous, or streamline flow and is composed of concentric cylindrical layers flowing past each other in a manner determined by the viscosity of the fluid. The fluid particles are retained in layers, and their motion is along parallel paths. As the flow rate is increased by changing the valve setting, it is found that the dye line will remain straight until a velocity is reached that causes it to waver and break into diffused patterns. The velocity at which the dye streak wavers and breaks is known as the critical velocity. At velocities greater than the critical velocity the colored dye filament becomes completely diffused in the main body of the fluid a short distance downstream from its point of injection. At velocities greater than the critical velocity the flow is termed *turbulent*, and the particles have a random motion that is transverse to the main flow direction and that causes the particles to intermingle in a random manner. In laminar flow the velocity of the fluid is maximum at the pipe axis and decreases to zero at the wall of the pipe, while in turbulent flow the velocity distribution is more uniform across the pipe diameter, as shown in Figure 6.2. By taking the average velocity as the characteristic veloc-

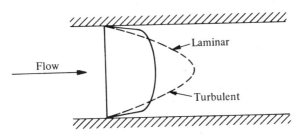

Figure 6.2. *Velocity profiles in a pipe.*

ity and the pipe diameter as the characteristic length, Reynolds showed that the character of the flow of a fluid in a pipe was dependent on the pipe diameter, the velocity of flow, the density of the fluid, and its viscosity. The combination of these four variables has already been seen in Chapter 5 to yield a dimensionless parameter known as the Reynolds number, $DV\rho/\mu$.

During the course of his experiments, Reynolds was able to obtain the change from laminar to turbulent flow at Reynolds numbers as low as 1200 and higher than 40,000. However, the conditions under which these high Reynolds numbers were obtained are not ordinarily found in commercial installations. These numbers at which the transition in character from laminar to turbulent flow occur are known as the Reynolds upper critical numbers and are not normally significant in the analysis of normal pipe flow. However, if the flow is initially turbulent and the velocity of the fluid is decreased, it will be found that all initial disturbances will disappear and that the flow will become laminar. The value of this Reynolds number, known as the Reynolds lower critical number, is generally agreed to be approximately 2000. The usual piping installation will be found to undergo a transition from laminar to turbulent flow at Reynolds numbers from 2000 to 4000, with the flow always laminar below a Reynolds number of 2000 and always turbulent above a Reynolds number of 4000. Between these two values there is a region known as the transition region in which the flow may be either laminar or turbulent. In the transition region a disturbance will cause the character of the flow to change from laminar to turbulent.

6.3 Laminar Flow in Tubes

Let us consider a horizontal circular tube in which an incompressible fluid is flowing steadily. Since the flow is steady, the fluid does not experience any acceleration, and we can draw a free-body diagram and impose the condition on the free body that it is in equilibrium; therefore the sum of the forces acting on the body must be zero. Consider the horizontal circular tube shown in Figure 6.3 and show all of the forces on the selected free body, i.e., the end

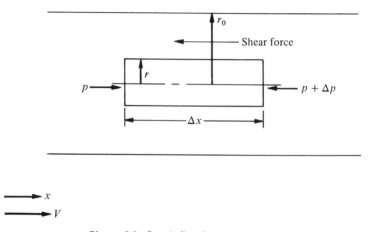

Figure 6.3. *Steady flow in a circular tube.*

forces and the shear forces over the surface. The direction of flow is to the right, and displacement will also be measured to the right. We shall denote the fact that there is a pressure gradient opposite to the direction of flow (a decrease from right to left) by letting the upstream pressure on the fluid element exceed the downstream pressure on the element by an amount Δp. Referring to Figure 6.3, we write the following equation for force equilibrium in the direction of flow:

$$p\pi r^2 - (p + \Delta p)\pi r^2 - 2\pi r (\Delta x)\tau = 0 \qquad (6.1)$$

where the last term on the left is the shear force over the outer periphery of the element. Dividing by $\pi r^2 (\Delta x)$, the volume of the element,

$$-\frac{\Delta p}{\Delta x}\frac{r}{2} = \tau \qquad (6.2)$$

Extending this derivation to the tube wall, we have

$$-\frac{\Delta p}{\Delta x}\frac{r_o}{2} = \tau_o \tag{6.3}$$

where τ_o is the shear stress at the wall. Since r_o and $\Delta p/\Delta x$ can be measured in the laboratory, equation (6.3) affords us the means to evaluate the shear stress at the wall of a circular pipe for the case of steady incompressible flow regardless of the character of the flow. Equations (6.2) and (6.3) are further interesting in that they were derived without knowing whether the fluid was flowing in either the laminar or turbulent flow regimes. Also, we conclude that the shear stress is zero at the center of the tube, increasing linearly to its maximum value at the wall of the tube, and is independent of the position along the tube.

It will be recalled from Section 1.6 that the viscosity and shear stress are related since shear stress is equal to the viscosity multiplied by the relative velocity between adjacent fluid layers and divided by the distance separating the layers. For the case of a cylinder this relation can be written as

$$\tau = -\mu\frac{\Delta V}{\Delta r} \tag{6.4}$$

where the minus sign is introduced to account for the fact that $\Delta V/\Delta r$ is negative. If equation (6.4) is substituted into equation (6.2),

$$\frac{\Delta V}{\Delta r} = \frac{1}{\mu}\frac{\Delta p}{\Delta x}\frac{r}{2} \tag{6.5}$$

Since $\Delta p/\Delta x$ represents the pressure drop per unit length of tube, it is independent of the radius r for a given tube. Thus,

$$\Delta V = \frac{1}{2}\frac{1}{\mu}\frac{\Delta p}{\Delta x} r\,\Delta r \tag{6.6}$$

To solve equation (6.6) it is necessary to obtain the summation of all of the $r\,\Delta r$ terms from the outside of the tube to any position r. Referring to Figure 6.4, it will be seen that this term is simply the area under a line sloped at 45 deg on a plot of r against r between the limits of r_o at the outside to r at any position. Therefore the summation of the $r\,\Delta r$ terms is $\frac{1}{2}r_o^2 - \frac{1}{2}r^2$. Corresponding to r_o, the outer radius of the tube, we have zero velocity. Therefore the solution to equation (6.6) becomes,

$$-V = \frac{\Delta p}{\Delta x}\frac{1}{4\mu}(r_o^2 - r^2) \tag{6.7}$$

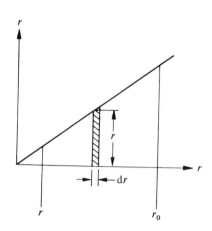

$$\sum_{r}^{r_0} r\Delta r = \frac{1}{2}{r_0}^2 - \frac{1}{2}r^2$$

where $\displaystyle\sum_{r}^{r_0}$ denotes the summation

between r_0 and r.

Figure 6.4. *Evaluation of* $r\,\Delta r$.

where the minus sign is used since $\Delta p/\Delta x$ is negative (pressure decreases in the direction of flow).

Equation (6.7) is the equation of a parabola with the maximum velocity along the center line of the tube. It is sometimes desired to know the average velocity that exists in a circular tube with the fluid flowing in the laminar flow regime. This average velocity can be determined by defining the average velocity as that velocity when multiplied by the area of the tube yields the volume flowing per unit time in the tube. Denoting $\bar{V}$ as the average velocity and Q as the volume rate of flow,

$$\bar{V}(A) = Q \tag{6.8a}$$

and

$$\bar{V} = \frac{Q}{A} \tag{6.8b}$$

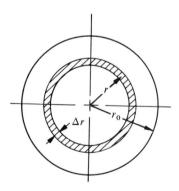

Figure 6.5. *Element of constant velocity in a circular tube.*

The quantity Q is the sum of the velocity in the tube at any radius multiplied by the area of the tube for which the velocity can be taken to be essentially constant. Figure 6.5 shows a ring of thickness Δr at radius r over which we can assume the velocity to be constant. Thus,

$$Q = \sum_{r}^{r_0} 2\pi r(\Delta r)(V) \tag{6.9}$$

Substituting for V,

$$Q = -\sum_{r}^{r_o} 2\pi r\, \Delta r \left(\frac{\Delta p}{\Delta x}\right)\frac{1}{4\mu}(r_o^2 - r^2) \tag{6.10}$$

and

$$Q = \left(\sum_{r}^{r_o} 2\pi r\, \Delta r\, \frac{\Delta p}{\Delta x}\,\frac{1}{4\mu}r_o^2 - \sum_{r}^{r_o} 2\pi r\, \Delta r\, \frac{\Delta p}{\Delta x}\,\frac{1}{4\mu}r^2\right) \tag{6.11}$$

The first summation on the right involves the term $r\,\Delta r$ and the second summation involves the term $r^3\,\Delta r$ since all other terms are not functions of r. Using the approach shown in Figure 6.4, we have for the summation $r\,\Delta r$, $\frac{1}{2}r_o^2 - \frac{1}{2}r^2$. For the $r^3\,\Delta r$ summation we plot r^3 against r, as shown in Figure 6.6. The area under this curve between r_o and r is $(r_o^4/4) - (r^4/4)$. Substituting

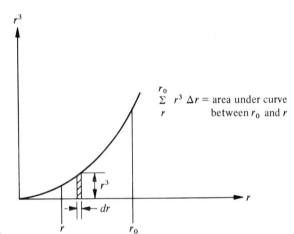

$$\sum_{r}^{r_0} r^3\, \Delta r = \text{area under curve between } r_0 \text{ and } r$$

Figure 6.6. *Evaluation of $r^3\,\Delta r$.*

both summations into (6.11) and simplifying, we have for the entire tube from $r = 0$ to $r = r_o$,

$$Q = -\frac{\Delta p}{\Delta x}\frac{\pi r_o^4}{8\mu} \tag{6.12}$$

and

$$\Delta p = -\frac{128\mu(\Delta x)Q}{\pi D_o^4} \tag{6.13}$$

Equation (6.13) is known as the Hagen-Poiseuille law, and it has been found that it checks very accurately with experimental results. It is most important to note that the pressure drop in laminar flow is independent of the character (roughness) of the pipe wall. Also, because of the excellent agreement between measured and calculated results in laminar flow, the Hagen-Poiseuille law has been used as the basis for measuring the viscosity of fluids in commercial visco-simeters.

Returning to equation (6.8),

$$\bar{V} = \frac{Q}{A} = \frac{\frac{\Delta p}{\Delta x}\frac{\pi r_o^4}{8\mu}}{\pi r_o^2} = -\frac{\Delta p}{\Delta x}\frac{r_o^2}{8\mu} \tag{6.14}$$

Comparing equation (6.14) with equation (6.7) with $r = 0$, we obtain the result that the average velocity in a circular tube for an incompressible fluid flowing in the laminar flow regime is one half of the maximum velocity.

ILLUSTRATIVE PROBLEM 6.1. At what radius from the center of a pipe does the velocity in laminar flow equal the average velocity?

Solution. From equation (6.7),

$$-V = \frac{\Delta p}{\Delta x}\frac{1}{4\mu}(r_o^2 - r^2)$$

and from equation (6.14),

$$-\bar{V} = \frac{\Delta p}{\Delta x}\frac{r_o^2}{8\mu}$$

Equating these expressions yields,

$$\frac{r_o^2}{2} = r_o^2 - r^2$$

and

$$r^2 = \frac{r_o^2}{2}$$

Therefore,

$$r = \frac{r_o}{\sqrt{2}}$$

Thus the radius at which the average velocity equals the local velocity is 70.7 percent of the tube radius.

For noncircular sections equation (6.1) can be written as,

$$[p - (p + dp)]A = P(\Delta x)\tau_o \qquad (6.15)$$

where A is the cross-sectional area of the flow channel and P the "wetted" perimeter of the channel. Simplifying and solving for the shear stress at the wall of the noncircular section, we have

$$\tau_o = -\frac{A}{P}\left(\frac{\Delta p}{\Delta x}\right) \qquad (6.16)$$

However, from equation (6.3) the shear stress for the circular channel is

$$\tau_o = -\frac{D_o}{4}\left(\frac{\Delta p}{\Delta x}\right) \qquad (6.17)$$

If we now equate the shear stresses given by equations (6.16) and (6.17), we can arrive at an "equivalent" diameter for the noncircular channel:

$$D_{eq} = \frac{4A}{P} \qquad (6.18)$$

D_{eq} is commonly known as the equivalent or hydraulic diameter of a noncircular pipe. Care must be exercised in using equation (6.18) for annular or other shapes that widely deviate from being nearly circular. However, we note that the derivation leading to equation (6.18) is independent of the character of the pipe flow and that it should apply whether the flow is laminar or turbulent.

ILLUSTRATIVE PROBLEM 6.2. A pipe is in the shape of a square of side L, an equilateral triangle of side L, and a rectangle whose width is half of its depth. Determine the equivalent diameter of each of these pipes.

Solution

1. Square:

$$A = L^2, \qquad P = 4L$$
$$D_{eq} = \frac{4A}{P} = \frac{4(L^2)}{4L} = L$$

The equivalent diameter of a square is simply the dimension of any side.

2. Equilateral triangle:

$$A = \frac{\sqrt{3}}{4} L^2, \qquad P = 3L$$

$$D_{eq} = \frac{4A}{p} = \frac{4\sqrt{3} L^2}{4(3L)} = \frac{L}{\sqrt{3}}$$

3. Rectangle:

$$A = \frac{L^2}{2}, \qquad P = \frac{L}{2} + \frac{L}{2} + L + L = 3L$$

$$D_{eq} = \frac{4A}{p} = \frac{4\frac{L^2}{2}}{3L} = \frac{2}{3}L$$

6.3a Stokes' Law

If a sphere is totally submerged in a fluid flowing past it with a relative velocity V, it can be shown that the force exerted on the sphere due to this relative motion arises predominantly from viscous effects on the surface of the sphere for Reynolds numbers equal to or less than unity. The resultant force on the sphere is given by equation (6.19), which was originally derived by Sir G. G. Stokes in 1851:

$$F = 6\pi r_o \mu V \tag{6.19}$$

where r_o is the outside radius of the sphere, μ the viscosity of the fluid, and V the relative velocity between the sphere and the undisturbed fluid.

If a spherical object falls freely in a still viscous fluid, it will reach a velocity relative to the fluid that remains constant and is known as the terminal velocity. At this time we can consider the body to be in equilibrium and treat it as a problem in fluid statics where the body is subjected to three forces, namely, its weight, the buoyant force due to displaced fluid, and the force due to the relative motion of the sphere (drag) and the fluid. These forces are shown in Figure 6.7.

Mathematically,

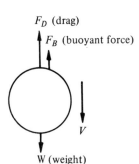

F_D (drag)

F_B (buoyant force)

V

W (weight)

Figure 6.7. *Free-body diagram of a freely falling sphere.*

$$F_{(drag)} + F_{(buoyant\ force)} = W \tag{6.20}$$

and

$$6\pi r_o \mu V + \tfrac{4}{3}\pi r_o^3 \gamma_F = \tfrac{4}{3}\pi r_o^3 \gamma_S \qquad (6.21)$$

where γ_S is the specific weight of the sphere and γ_F the specific weight of the fluid. Simplifying,

$$V = \frac{2}{9}\frac{r_o^2}{\mu}(\gamma_S - \gamma_F) \qquad (6.22)$$

Equation (6.22) is known as Stokes' Law, and it is frequently used in problems for the determination of small particle sizes. It is also useful as a method of obtaining the viscosity of a fluid by measuring the terminal velocity of a falling sphere either optically, mechanically, or electrically.

ILLUSTRATIVE PROBLEM 6.3. A spherical metal ball is allowed to fall freely in 68°F water. If the ball is made of aluminum that weighs 168.8 lb/ft³ and is 0.01 in. in diameter, determine its terminal velocity.

Solution. At 68°F water has a viscosity of 1 cP. From illustrative problem 1.5, we have 1 P = 0.00209 lb sec/ft². Therefore 1 cP = 2.09×10^{-5} lb sec/ft². Using 62.4 for the specific weight of water and $r_o = 0.01/(2 \times 12) = 0.000417$ ft,

$$V = \frac{2}{9}\frac{r_o^2}{\mu}(\gamma_S - \gamma_F)$$

Therefore,

$$V = \frac{2}{9}\frac{(0.000417)^2}{2.09 \times 10^{-5}}(168.8 - 62.4)$$

and

$$V = 0.197 \text{ ft/sec}$$

6.4 Boundary Layer[1]

The parabolic velocity profile of fluid flowing in laminar flow inside a pipe is due partly to the viscosity of the fluid and partly to the adhesive force between the liquid and the pipe wall. As the flow remains laminar, the velocity profile

[1]The material in this section is developed with permission of the Van Nostrand Reinhold Company from *Engineering Heat Transfer* by S. T. Hsu, Litton Educational Publishing, Inc., New York, 1963, pp. 209–211.

remains parabolic. When the flow becomes turbulent, the velocity decreases only very slightly from the axis of the pipe to the vicinity of the pipe wall. However, there is a steep velocity gradient in the thin layer of liquid next to the stationary boundary. This is called the boundary layer. Although the main body of fluid is turbulent, there still exists a thin laminar layer immediately next to the pipe wall. Some typical velocity distributions are shown in Figure 6.8.

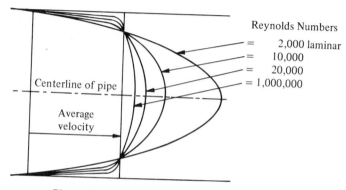

Figure 6.8. Velocity distributions in a smooth pipe.

Let us consider the flow of fluid over a flat plate held parallel to the direction of flow (see Figure 6.9). The vertical scale is purposely enlarged in order to show the detail of the flow pattern. When the fluid passes the leading edge of the plate, the velocity gradient and the viscous boundary shear are high. The fluid is moving in the laminar state, and the boundary layer is thin. This is called the laminar boundary layer. As the fluid travels farther down the stream along the plate, the retardation of fluid flow increases due to shearing force, and the boundary layer also grows in thickness. As a result the velocity gradient gradually decreases and concurrently the boundary shear is reduced as the thickness increases. When the boundary layer becomes thick enough, the particles begin to move out of the smooth layers or laminae, the laminar motion

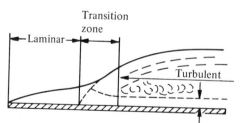

Laminar sublayer   Figure 6.9. Boundary-layer transition.

becomes unstable, and finally the flow becomes turbulent. However, under the turbulent boundary layer there is still a thin layer of fluid immediately next to the solid boundary, and this thin layer is flowing in the laminar regime. This layer is called the laminar sublayer of the turbulent boundary layer. The layer of transition from the laminar sublayer to the turbulent layer is called the buffer layer. Once the boundary layer is in the turbulent state, velocity distribution becomes more nearly constant in a lateral direction. A laminar boundary layer cannot change suddenly into a turbulent one, and a transition zone is found to exist between the turbulent and laminar zones. The turbulent motion is accompanied by an increase of the boundary shear and by an expansion of the thickness of the boundary layers. For a submerged plate, the transition takes place in the range of Reynolds numbers from 500,000 to 1 million, where the characteristic dimension is the distance from the leading edge of the plate. The number depends on the initial state of flow and also on the shape of the front edge and the roughness of the plate.

The thickness of the boundary layer is defined as the distance from the boundary to the point where the velocity reaches the value of the main stream velocity. It is extremely difficult actually to measure the thickness of the boundary layer, especially in turbulent flow. But the thickness of the boundary layer can be predicted by analytical methods or from the experimental results of velocity and temperature measurements. In laminar boundary layers the velocity profiles join the outside velocity curve asymptotically.

The velocity profile inside the boundary layer may be approximated by an analytical method based upon the theory of velocity distribution in pipes, and the drag force due to the boundary layer may also be derived by an analytical method. A complete discussion of these methods, however, is too advanced to fall within the scope of this book, and the student is referred to the references at the end of this chapter for further information on this subject.

6.5 Pressure Losses in Pipe Flow

6.5a Friction

In Chapter 4 it was indicated that the Bernoulli equation can and has been used when a real fluid flows in a pipe. It will be recalled that a term was added to the right side of the equation to account for "head losses" that occur in actual flow situations. Specifically we wrote

$$\frac{p_1}{\gamma_1} + \frac{V_1^2}{2g} + Z_1 = \frac{p_2}{\gamma_2} + \frac{V_2^2}{2g} + Z_2 + \text{Losses}_{1 \to 2} \qquad (6.23)$$

For a horizontal pipe of constant diameter,

$$\text{Losses}_{1\rightarrow2} = \frac{p_1 - p_2}{\gamma} \qquad (6.24)$$

namely, the loss in head between two sections of a horizontal pipe in which an incompressible fluid is flowing is simply the difference in the static pressures that exist at each of the sections.

When an incompressible fluid flows in a pipe and the flow is turbulent, it has been found experimentally that the head loss is a function of the length of the pipe, the diameter of the pipe, the surface roughness of the pipe wall, the velocity of the fluid, the density of the fluid, and the viscosity of the fluid. Using the techniques of dimensional analysis we can arrive at a functional relationship for the head loss when any incompressible fluid flows steadily in a pipe. It will be noted that this problem is illustrative problem 5.3 and that the solution can be written in the following form:

$$\text{Losses}_{1\rightarrow2} = \text{Constant} \frac{V^2}{2g}\left(\frac{DV\rho}{\mu}\right)^{-b}\left(\frac{V^2}{gD}\right)^{-f}\left(\frac{e}{D}\right)^{e}\left(\frac{L}{D}\right)^{g} \qquad (6.25)$$

Since experiments have shown that the Froude number V^2/gD is not important in this type of flow and that the losses vary directly with L/D, we can rewrite (6.25) as follows:

$$\text{Losses}_{1\rightarrow2} = \text{Constant} \frac{LV^2}{D2g}\left(\frac{DV\rho}{\mu}\right)^{-b}\left(\frac{e}{D}\right)^{e} \qquad (6.26)$$

The form of equation most generally used to calculate the frictional losses in pipes is known as the Darcy-Weisbach equation:

$$h_f = f\frac{L}{D}\frac{V^2}{2g} \qquad (6.27)$$

where h_f is the head loss in feet of fluid flowing and f is known as the friction factor. If we compare equations (6.27) and (6.26), we conclude that the friction factor in pipe flow is a function of the Reynolds number and e/D. The ratio of e/D is commonly called the relative roughness, and e is called the absolute roughness of the pipe wall.

Earlier in this chapter we derived the Hagen-Poiseuille equation for the pressure drop for an incompressible fluid flowing in laminar flow inside of a pipe. Rewriting this equation and noting that $Q = AV$,

$$\Delta p = \gamma h_f = \frac{128\mu LQ}{\pi D_o^4} = \frac{128\mu L \frac{\pi D_o^2}{4} V}{\pi D_o^4} \qquad (6.28)$$

Therefore,

$$h_f = f\frac{L}{D_o}\frac{V^2}{2g} = \frac{128\mu L \frac{\pi D_o^2}{4} V}{\pi D_o^4} \qquad (6.29)$$

Substituting $\gamma = \rho g$ and simplifying,

$$f = \frac{64}{\dfrac{D_o V \rho}{\mu}} = \frac{64}{Re} \qquad (6.30)$$

where Re is based upon the *average* velocity. We therefore conclude that the friction factor in laminar flow is simply 64 divided by the Reynolds number and is independent of the relative roughness of the pipe.

Unfortunately, in the turbulent flow regime it is not possible to obtain an analytical solution for the friction factor as we just did for the case of laminar flow. Most of the usable data available for evaluating the friction factor in turbulent flow has been derived from experiments. Prior to 1933 most of these results were scattered throughout the technical literature. In 1933, J. Nikuradse published his work on pipes whose walls had been artificially roughened using glued sand grains of different diameters. He termed the diameter of the sand grains (e) the absolute roughness and the ratio of the sand grain diameter to the inside pipe diameter (e/D) the relative roughness. The results of Nikuradse's work is shown graphically in Figure 6.10. In the laminar zone ($Re \leqslant \sim 2100$) the friction factor is independent of the relative or absolute roughness. In the turbulent zone the friction factor is seen to be a function of the relative roughness and the Reynolds number, and it will be seen that as the Reynolds number increases the friction factor becomes constant for each given value of the relative roughness and consequently independent of the Reynolds number.

In the same year, 1933, R. J. S. Pigott and E. Kemler published papers in which much of the existing data on friction in pipes were correlated and plotted in a diagram similar to Figure 6.10. This type of diagram in which the logarithm of the friction factor is plotted as a function of the logarithm of the Reynolds number is known as a Stanton diagram and is a very convenient portrayal of the data. In 1944 L. F. Moody published his paper on friction factors in pipe

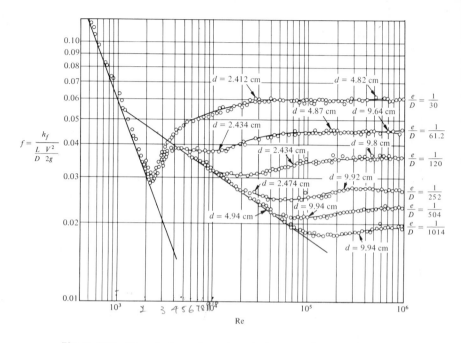

Figure 6.10. *Nikuradse's sand-roughened pipe tests. Reproduced by permission of the Van Nostrand Reinhold Company from* Engineering Heat Transfer *by S. T. Hsu, Litton Educational Publishing, Inc., 1963, p. 213.*

flow and in this paper gave his results in the form of a Stanton diagram, which has proved to be the most widely used source of friction factors for new or clean commercial pipes. This figure is reproduced as Figure 6.11. Qualitatively the trends in Nikuradse's data shown in Figure 6.10 and the Moody diagram shown in Figure 6.11 are in good agreement. Detailed differences exist in the critical and transition zones as might be expected when comparing artificially roughened pipes with commercial pipes. Also, the values of the absolute roughness of pipe used by Moody and given in Table 6.1 are arbitrary and not comparable to the sand grain diameters used by Nikuradse. The solid line on the Moody diagram indicates the value of the Reynolds number at which the flow becomes independent of Reynolds number and becomes a function of e/D only. This is noted to be the zone of complete turbulence, rough pipes. Figure 6.12 gives the values of e/D for various types of pipe and also the friction factor for complete turbulence, rough pipes. It will be noted that the friction factor in the turbulent zone decreases with increasing Reynolds number until the limiting value of complete turbulence, rough pipes, is reached. For convenience

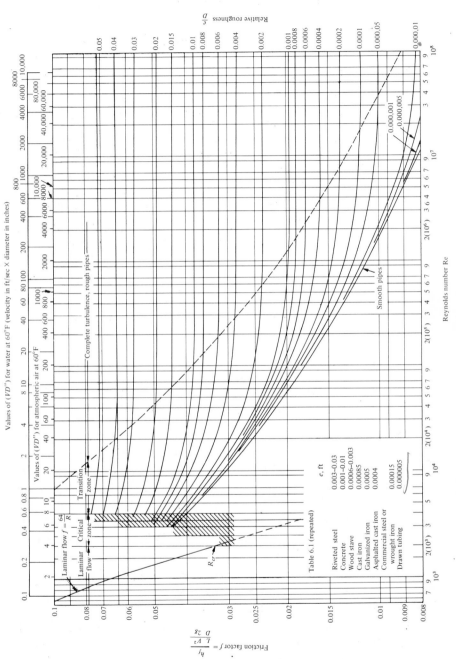

Figure 6.11. *Friction factors for any kind and size of pipe. Reproduced with permission from Pipe Friction Manual, 3rd ed., Hydraulic Institute, New York, 1961.*

177

Table 6.1 Absolute Roughness of Pipes

Material	e (ft)
Glass, new commercial pipe surfaces, drawn tubing (brass, copper, lead)	0.000005
Commercial steel or wrought iron	0.00015
Asphalted cast iron	0.0004
Galvanized iron	0.0005
Cast iron	0.00085
Wood stave	0.0006-0.0003
Concrete	0.001-0.01
Riveted steel	0.003-0.03

Figure 6.13 is included since it is a convenient presentation of both the kinematic viscosity and Reynolds number for many liquids and gases.

ILLUSTRATIVE PROBLEM 6.4. Water at 50°F is supplied from a reservoir to a 1000-ft-long, horizontal, round, concrete pipe 36 in. in diameter. Determine the frictional pressure drop in the pipe if the flow is 20,000 gal/min. Neglect minor loses.

Solution. Using 231 in^3. in 1 gal.

$$Q = \left(\frac{20,000}{60}\right)\left(\frac{231}{1728}\right) = 44.6 \text{ ft}^3/\text{sec}$$

Since $Q = AV$,

$$V = \frac{Q}{A} = \frac{(44.6)}{\dfrac{\pi}{4}\dfrac{(36)^2}{144}} = 6.3 \text{ ft/sec}$$

The product of (V) (D) is $6.3 \times 36 = 227$. Using Figure 6.13, we can evaluate the Reynolds number directly:

$$Re \simeq 1.6 \times 10^6$$

The flow is therefore turbulent. The student should check the value of Re by direct calculation of $DV\rho/\mu$ or $DV\gamma/\mu g$.

From Table 6.1, e for concrete varies from 0.001 to 0.01 ft. Using the arithmetic average of 0.005,

$$\frac{e}{D} = \frac{0.005 \times 12}{36} = 0.0017$$

From Figure 6.11, $f = 0.023$.

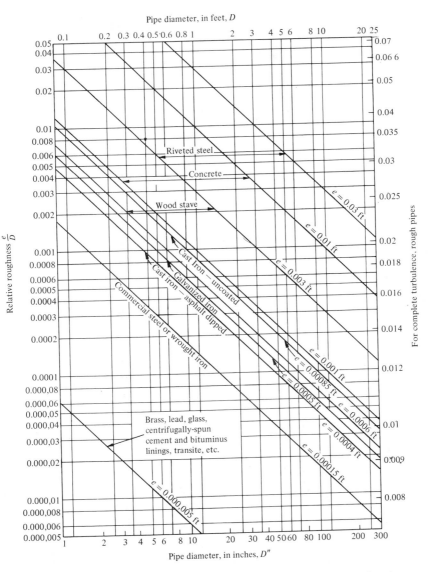

Figure 6.12. *Relative roughness factors for new clean pipes. Reproduced with permission from* Pipe Friction Manual, *3rd ed., Hydraulic Institute, New York, 1961.*

Figure 6.13. *Kinematic viscosity and Reynolds number chart.* Reproduced with permission from Pipe Friction Manual, 3rd ed., Hydraulic Institute, New York, 1961.

The pressure drop due to friction is

$$h_f = f \frac{L}{D} \frac{V^2}{2g} = 0.023 \left(\frac{1000}{\frac{36}{12}} \right) \frac{(6.3)^2}{2g} = 4.72 \text{ ft of water}$$

or

$$h_f = 4.72 \times \frac{62.4}{144} = 2.04 \text{ psi}$$

ILLUSTRATIVE PROBLEM 6.5. Water flows in a steel pipe whose inside diameter is 2.067 in. If the pipe is 200 ft long and a pump can furnish a head of 50 ft to overcome friction losses, determine the flow in the pipe.

Solution. In this problem it is not possible to obtain a direct solution, and it will be necessary to use successive iterations (trial and error). The procedure is to assume a value of friction factor and to solve the problem based upon this assumption. The friction factor is then evaluated using the solution based upon the initial assumption. If the calculated and assumed values agree, this is the solution. If they do not agree, a new assumption is made and the procedure is repeated.

Assume $f = 0.02$.

Therefore,

$$50 = 0.02 \frac{200}{\frac{2.067}{12}} \frac{V^2}{2g}$$

and

$$V = 11.8 \text{ ft/sec}$$

$$VD = 11.8 \times 2.067 = 24.4$$

From Figure 6.11, $Re \simeq 2.0 \times 10^5$. From Table 6.1, $e = 0.00015$ ft. Therefore,

$$\frac{e}{D} = \frac{0.00015}{\frac{2.067}{12}} = 0.00087$$

and from Figure 6.11, $f = 0.021$

The calculated value of f based upon an assumed value of f of 0.020 is 0.021. This is satisfactory and the value of 11.8 ft/sec will be used. Therefore,

$$Q = AV = 11.8 \times \frac{\pi}{4} \frac{(2.067)^2}{144} = 0.277 \text{ ft}^3/\text{sec}$$

$$\frac{0.277 \times 60 \times 1728}{231} = 124 \text{ gal/min}$$

While Figure 6.11 is convenient when solving problems in pipe friction, it is often desirable to have an explicit formula expressing friction factor as a function of the relative roughness and the Reynolds number. In 1947 Moody published such a formulation, which is sufficiently accurate for most engineering purposes, namely,

$$f \simeq 0.0055 \left[1 + \left(20,000 \frac{e}{D} + \frac{10^6}{Re} \right)^{1/3} \right] \tag{6.31}$$

Equation (6.31) is cumbersome and very often the friction factor is expressed as

$$f = \frac{a}{(Re)^n} \tag{6.32}$$

where a and n depend on the relative roughness and the Reynolds number. Values for a and n can be found in *Thermodynamics of Fluid Flow* (reference 11, pp. 30–31).

The problem of evaluating friction factors for old pipes and allowing for the deterioration of new pipes is particularly difficult since pipe deterioration is a function of the fluid and pipe chemical properties. Thus it is recommended that the Moody curve and the data presented in this chapter should not be used when estimating the pressure losses in old pipes or to allow for the aging of new pipes. If at all possible, experience with an existing installation will yield the best data for estimating the pressure drop in old or aging pipes. Data in the engineering literature are not satisfactory and may yield results that are in large error when applied to a specific situation.

6.5b Other Losses (Minor Losses)

In addition to the pressure drop incurred due to friction as a fluid flows in a pipe, other losses in pressure occur. These other losses occur at the entrance to a pipe, at abrupt changes in section in a pipe, at valves, at fittings, at bends, and at the pipe exit; they are often called minor losses. The data available in the literature on the losses in values and fittings have been presented in two different forms. The first form expresses the fact that these losses can be expressed in terms of velocity heads of fluid flowing; i.e.,

$$h = K \frac{V^2}{2g} \tag{6.33}$$

The value of K increases with increasing roughness and decreases with increasing Reynolds numbers but primarily depends on the geometric shape of the valve or fitting. Figure 6.14 gives the resistance coefficients for pipe fittings and Figure 6.15 gives the resistance coefficients for valves, couplings, and unions. To appreciate that these values are approximate and to appreciate the wide variations that are found in specific fittings, Table 6.2 has been included. It is apparent from this table that, at best, the calculation of these losses is approximate and that the values given in Figures 6.14 and 6.15 for the resistance coefficients are just good approximations.

The second method of presenting these losses is to express them as equivalent lengths of pipe L_{eq} that have the same head loss for the same discharge. Therefore,

$$f \frac{L_{eq}}{D} \frac{V^2}{2g} = K \frac{V^2}{2g} \tag{6.34}$$

and

$$L_{eq} = \frac{KD}{f} \tag{6.35}$$

These minor losses are in excess of the friction loss produced in a straight pipe whose length is equal to the length of the axis of the pipe (see Table 6.3).

ILLUSTRATIVE PROBLEM 6.6. A commercial steel pipeline has an inside diameter of 6 in. and the flow is completely turbulent. If the sum of all the loss coefficients in the pipe (K) is 12, determine the equivalent length that must be added to the actual pipe length in order to cause the same resistance to flow.

Solution. For steel pipe, $e = 0.00015$ and $e/D = 0.00015/(\frac{6}{12}) = 0.00030$. For completely turbulent flow, $f = 0.015$. Therefore,

$$L_{eq} = \frac{12 \times \frac{6}{12}}{0.015} = 400 \text{ ft to be added}$$

In addition to valves and fittings, most pipelines have abrupt changes at entrances, exits, reducers, increasers, diffusers, and bends. One of these losses,

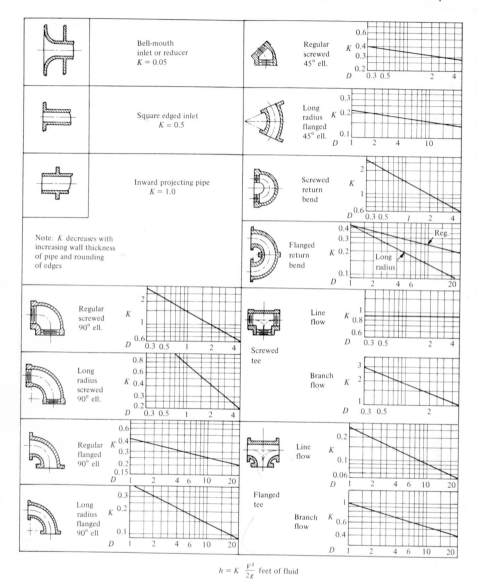

$$h = K \frac{V^2}{2g} \text{ feet of fluid}$$

Figure 6.14. *Resistance coefficients for valves and fittings. The value of D is nominal iron pipe size. For velocities below 15 ft/sec, check valves and foot valves will be only partially open and will exhibit higher values of K than that shown. Reproduced with permission from* Pipe Friction Manual, *3rd Ed., Hydraulic Institute, New York., 1961.*

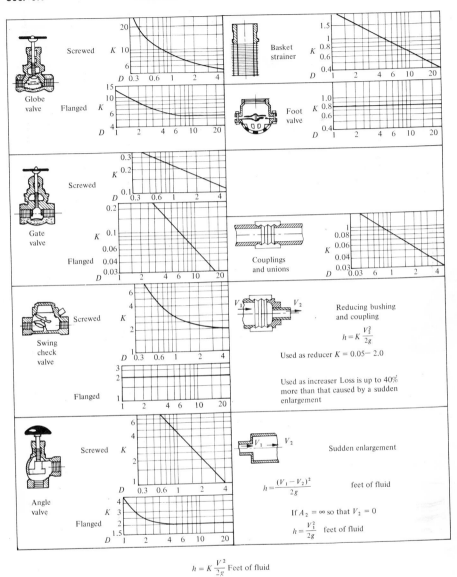

$$h = K \frac{V^2}{2g} \text{ Feet of fluid}$$

Figure 6.15. *Resistance coefficients for valves and fittings. See Figure 6.14. Reproduced with permission from* Pipe Friction Manual, *3rd ed., Hydraulic Institute, New York, 1961.*

Table 6.2† Resistance Coefficients for Valves and Fittings

Fitting	Approximate Range of Variation for K	
		Range of Variation
90-deg elbow	Regular screwed	±20% above 2-in. size
	Regular screwed	±40% below 2-in. size
	Long radius, screwed	±25%
	Regular flanged	±35%
	Long radius, flanged	±30%
45-deg elbow	Regular screwed	±10%
	Long radius, flange	±10%
180-deg bend	Regular screwed	±25%
	Regular flanged	±35%
	Long radius, flanged	±30%
Tee	Screwed, line, or branch flow	±25%
	Flanged, line, or branch flow	±35%
Globe valve	Screwed	±25%
	Flanged	±25%
Gate valve	Screwed	±25%
	Flanged	±50%
Check valve	Screwed	±30%
	Flanged	$\begin{cases} +200\% \\ -80\% \end{cases}$
Sleeve check valve		Multiply flanged values by 0.2–0.5
Tilting check valve		Multiply flanged values by 0.13–0.19
Drainage gate check		Multiply flanged values by 0.03–0.07
Angle valve	Screwed	±20%
	Flanged	±50%
Basket strainer		±50%
Foot valve		±50%
Couplings		±50%
Unions		±50%
Reducers		±50%

†Reproduced with permission from *Pipe Friction Manual*, 3rd ed., Hydraulic Institute, New York, 1961.

Table 6.3† Equivalent Length of New Straight Pipe for Valves and Fittings for Turbulent Flow Only

Fittings			¼	⅜	½	¾	1	1¼	1½	2	2½	3	4	5	6	8	10	12	14	16	18	20	21
Regular 90° ell	Screwed	Steel	2.3	3.1	3.6	4.4	5.2	6.6	7.4	8.5	9.3	11	13										
		Castiron										9.0	11										
	Flanged	Steel			0.92	1.2	1.6	2.1	2.4	3.1	3.6	4.4	5.9	7.3	8.9	12	14	17	18	21	23	25	30
		Castiron										3.6	4.8		7.2	9.8	12	15	17	19	22	24	28
Long radius 90° ell	Flanged	Steel	1.5	2.0	1.1	1.3	1.6	2.0	2.3	2.7	2.9	3.4	4.2	5.0	5.7	7.0	8.0	9.0	9.4	10	11	12	14
		Castiron										2.9	3.4		4.7	5.7	6.8	7.8	8.6	9.6	11	11	13
Regular 45° ell	Screwed	Steel	0.34	0.52	0.71	0.92	1.3	1.7	2.1	2.7	3.2	4.0	5.5										
	Flanged	Steel			0.45	0.59	0.81	1.1	1.3	1.7	2.0	2.6	3.5	4.5	5.6	7.7	9.0	11	13	15	16	18	22
		Castiron										2.1	2.9		4.5	5.5	6.7	7.7	8.7	9.9?	12	17	20
Tee-line flow	Screwed	Steel	0.79	1.2	1.7	2.4	3.2	4.6	5.6	7.7	9.3	12	17										
	Flanged	Steel			0.69	0.82	1.0	1.3	1.5	1.8	1.9	2.2	2.8	3.3	3.8	4.7	5.2	6.0	6.4	7.2	7.6	8.2	9.6
		Castiron										1.9	2.2		3.1	3.9	4.6	5.2	5.9	6.5	7.2	7.7	8.8
Tee-branch flow	Screwed	Steel	2.4	3.5	4.2	5.3	6.6	8.7	9.9	12	13	17	21										
	Flanged	Steel			2.0	2.6	3.3	4.4	5.2	6.6	7.5	9.4	12	15	18	24	30	34	37	43	47	52	62
		Castiron										7.7	9.4		15	20	25	30	35	39	44	49	57
180° return bend	Regular flanged	Steel	2.3	3.1	0.92	1.2	1.6	2.1	2.4	3.1	3.6	4.4	5.9	7.3	8.9	12	14	17	18	21	23	25	30
		Castiron										3.6	4.8		7.2	9.8	12	15	17	19	22	24	28
	Long radius flanged	Steel			1.1	1.3	1.6	2.0	2.3	2.7	2.9	3.4	4.2	5.0	5.7	7.0	8.0	9.0	9.4	10	11	12	14
		Castiron										2.9	3.4		4.7	5.7	6.8	7.8	8.6	9.6	11	11	13
Globe valve	Screwed	Steel	21	22	22	24	29	37	42	54	62	79	110										
		Castiron										65	86										
	Flanged	Steel			38	40	45	54	59	70	77	94	120	150	190	260	310	390					
		Castiron										79	99		150	210	270	330					
Gate valve	Screwed	Steel	0.32	0.45	0.56	0.67	0.84	1.1	1.2	1.5	1.7	1.9	2.5										
		Castiron										1.6	2.0										
	Flanged	Steel								2.6	2.7	2.8	2.9	3.1	3.2	3.2	3.2	3.2	3.2	3.2	3.2	3.2	3.2
		Castiron										2.3	2.4		2.6	2.7	2.8	2.9	2.9	3.0	3.0	3.0	3.0
Angle valve	Screwed	Steel	12.8	15	15	15	17	18	18	21	22	28	38										
		Castiron										23	31										
	Flanged	Steel								18	18	18	31	50	63	90	120	140	160	190	210	240	300
		Castiron										15	31		52	74	98	120	150	170	200	230	280
Swing check valve	Screwed	Steel	7.2	7.3	8.0	8.8	11	13	15	17		27	38										
	Flanged	Steel			3.8	5.3	7.2	10	12	17	22	27	31		63	90	120	140	160	190	210		
		Castiron													52	74	98	120	150	170	200		
Coupling or union	Screwed	Steel	0.14	0.18	0.21	0.24	0.29	0.36	0.39	0.45	0.47	0.53	0.65										
		Castiron										0.44	0.52										
Bell mouth inlet		Steel	0.04	0.07	0.10	0.13	0.18	0.26	0.31	0.43	0.52	0.55	0.77	0.95	1.6	2.3	2.9	3.5	4.0	4.7	5.3	6.1	7.6
		Castiron										0.44			1.3	1.9	2.4	3.0	3.6	4.3	5.0	5.7	7.0
Square mouth inlet		Steel	0.44	0.68	0.96	1.3	1.8	2.6	3.1	4.3	5.2	6.7	9.5	13	16	23	29	35	40	43	53	61	76
		Castiron										5.5	7.7		13	19	24	30	36	47	50	57	70
Reentrant pipe		Steel	0.88	1.4	1.9	2.6	3.6	5.1	6.2	8.5	10	11	19	25	32	45	58	70	80	95	110	120	150
		Castiron										15			26	37	49	61	73	86	100	110	140
Sudden enlargement																							

$$h = \frac{(V_1 - V_2)^2}{2g} \text{ feet of liquid; if } V_2 = 0,\ h = \frac{V_1^2}{2g} \text{ feet of liquid}$$

†Reproduced with permission from *Pipe Friction Manual*, 3rd ed., Hydraulic Institute, New York, 1961.

the case of loss at a sudden enlargement, can be calculated, with the results agreeing reasonably well with experiment. When this is done, the head loss is found to be

$$h = \frac{(V_1 - V_2)^2}{2g} \tag{6.36}$$

where V_1 is the upstream velocity and V_2 the downstream velocity. Such a condition exists when a pipe exits into a large tank or reservoir. For this case equation (6.36) gives a loss equal to one velocity head in the pipe. Alternatively we can argue that the kinetic energy in the pipe is converted to internal energy and does not appear again as a pressure head. In this sense one velocity head is "lost" at the exit of the pipe. For a sudden contraction Figure 6.16 gives the

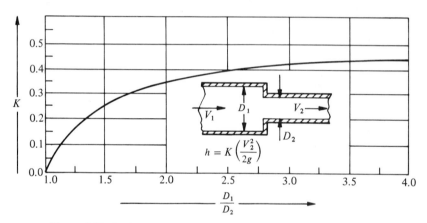

Figure 6.16. *Resistance coefficients for reducers. Reproduced with permission from* Pipe Friction Manual, *3rd ed., Hydraulic Institute, New York, 1961.*

loss coefficient. It will be noted that if this curve is applied to the sharp entrance of a pipe from a large reservoir, the loss coefficient is half a velocity head. As will be also seen from Figure 6.14, this loss can be greatly reduced by rounding off the entrance to the pipe.

In addition to the foregoing, Figures 6.17 and 6.18 can be used to obtain the resistance coefficient, K, for increasers, diffusers, and bends.

ILLUSTRATIVE PROBLEM 6.7. A large open reservoir is connected to a steel tube whose inside diameter is 2 in. A screwed gate valve is placed in the tube near its outlet, and the tube is 50 ft long. If the level of water in the tank is 50 ft above the outlet of the tube, determine the rate at which the water flows when the valve is opened. Assume that the loss coefficients for bends and elbows add up to 3.0.

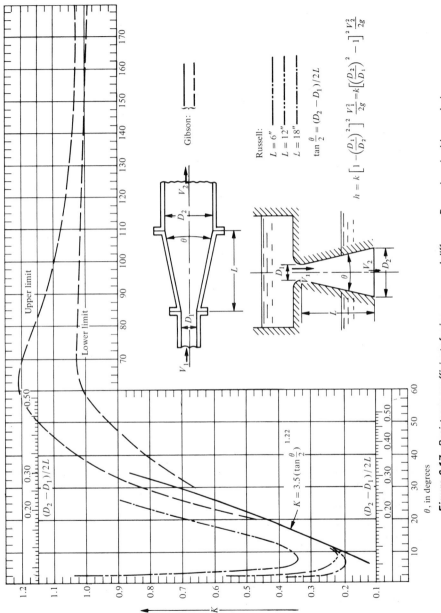

Figure 6.17. *Resistance coefficients for increasers and diffusers. Reproduced with permission from Pipe Friction Manual, 3rd ed., Hydraulic Institute, New York. 1961.*

189

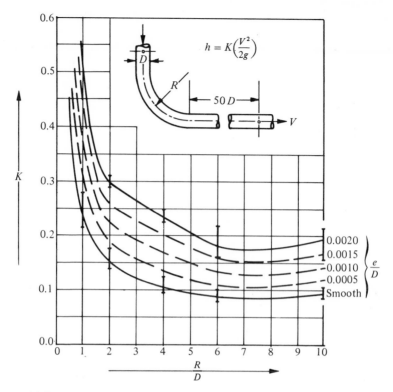

(a) Resistance coefficients for 90 degree bends of uniform diameter.

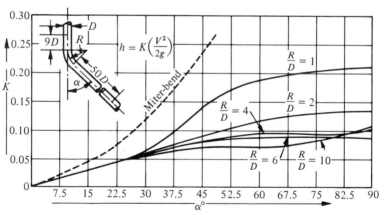

(b) Resistance coefficients for bends of uniform diameter and smooth surface at Reynolds No. $\sim 2.25 \times 10^3$

Figure 6.18. (a) Resistance coefficients for 90-deg bends of uniform diameter. (b) Resistance coefficients for bends of uniform diameter and smooth surface at Reynolds number 2.25×10^3. Reproduced with permission from Pipe Friction Manual, 3rd ed., Hydraulic Institute, New York, 1961.

Solution. Writing a Bernoulli equation between sections 1 and 2 in the accompanying figure,

$$\frac{p_1}{\gamma} + \frac{V_1^2}{2g} + Z_1 = \frac{p_2}{\gamma} + \frac{V_2^2}{2g} + Z_2 + \text{Losses}_{1 \to 2}$$

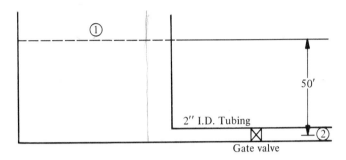

Since we have atmospheric pressure at sections 1 and 2 and V_1 can be taken to be zero,

$$\frac{V_2^2}{2g} + \text{Losses}_{1 \to 2} = Z_1 - Z_2$$

where we have already termed $V_2^2/2g$ to be the exit loss. Inserting an entrance loss, a loss term for the valve, and a friction loss term as well as the other losses, we have

$$\frac{V_2^2}{2g}\left[1 + f\frac{L}{D} + 3.0 + 0.5 + 0.17\right] = Z_1 - Z_2 = 50$$

In general, f is a function of the velocity V, and an explicit solution cannot be obtained for this problem. However, let us assume that the flow is fully turbulent:

$$\frac{e}{D} = \frac{0.00015}{\frac{2}{12}} = 0.0009$$

and f for fully developed turbulent flow is 0.019. Therefore,

$$\frac{V_2^2}{2g}\left[1 + \frac{0.019 \times 50}{\frac{2}{12}} + 3.0 + 0.5 + 0.17\right] = 50$$

and

$$V_2 = 17.7 \text{ ft/sec}$$

At this point it is necessary to check the Reynolds number to verify the assumption of fully turbulent flow. Since $VD = 17.7 \times 2 = 35.4$, $Re \simeq 2.0 \times 10^5$ from Figure 6.13. From Figure 6.11 we have $f = 0.021$. As a second trial assume $f = 0.021$, and solving we have $V_2 = 17.1$ ft/sec. For this velocity $Re \simeq (17.1/17.7) \times 2 \times 10^5 = 1.9 \times 10^5$ and $f = 0.021$. Therefore the solution is that water flows at the rate of 17.1 ft/sec in the pipe.

ILLUSTRATIVE PROBLEM 6.8. Air flows in a $\frac{1}{2}$-in. schedule 80 steel pipe (i.d. $= 0.546$ in.) from a large storage tank. The piping system will contain twelve 90-deg regular screwed elbows, two screwed gate valves, one screwed globe valve, three screwed tees with line flow, and 200 ft of straight pipe. If 11.3 lb/min flows and the air has a specific weight of 5.36 lb/ft^3, determine the pressure drop in the system. Assume that the tank pressure is 1000 psig, that air temperature is 1000°F, and that the pressure drop in the pipe is less than 10 percent of the tank pressure. If the pressure drop exceeds 10 percent of the tank pressure, it is incorrect to apply the equations for incompressible flow to this flow situation.

Solution. The velocity in the pipe is,

$$\dot{W} = \gamma A V; \qquad \frac{11.3}{60} = 5.36 \times \frac{\pi}{4} \frac{(0.546)^2}{144} \times V$$

and

$$V = \frac{11.3}{60} \frac{4}{\pi} \frac{144}{(0.546)^2 \, 5.36} = 21.5 \text{ ft/sec}$$

$$VD = 21.5 \times 0.546 = 11.8$$

From Figure 6.13, $Re \simeq 8 \times 10^3$ for air at 100 psia and 1000°F. Since the viscosity of a gas is primarily dependent on its temperature, the viscosity at 1014.7 psia and 1000°F will be the same as at 100 psia and 1000°F. The only term in the Reynolds number that changes due to pressure is the specific weight. Therefore the specific weight (and consequently the Reynolds number) at 1014.7 psia will exceed the value at 100 psia by the ratio of 1014.7:100. For the conditions of this problem,

$$Re \simeq 8 \times 10^3 \times \frac{1014.7}{100} \simeq 8 \times 10^4$$

For commercial steel pipe,

$$e = 0.00015 \text{ ft}$$

Therefore,

$$\frac{e}{D} = \frac{0.00015}{\dfrac{0.546}{12}} = 0.0033$$

From Figure 6.11, $f = 0.028$.

Expressing the friction loss in terms of velocity heads yields

$$K_f = f\frac{L}{D}$$

or

$$K_f = 0.028\,\frac{200}{\dfrac{0.546}{12}} = 123.0$$

For twelve 90-deg regular screwed elbows in a $\frac{1}{2}$-in. pipe size, $K = 2$ for each elbow.

For two screwed gate valves, $K = 0.32$ per valve.

For one screwed globe valve, $K = 14$.

For three screwed tees with line flow, $K = 0.9/$tee.

For entrance, $K = 0.5$.

For exit, $K = 1$.

Adding all of these terms we have

$$K = 123.0 + 12(2) + 2(0.32) + 14 + 3(0.9) + 1 + 0.5 = 165.8$$

Therefore,

$$h_f = 165.8\left[\frac{(21.5)^2}{2g}\right]\frac{5.36}{144} = 44.2 \text{ psi}$$

Since this figure is within the 10 percent limit the assumption of incompressible flow is valid for this problem.

Before continuing further, two cautions must be noted at this time. The tables and charts in this chapter for the calculation of friction and minor losses are based upon nominal IPS (iron pipe size). In all of the illustrative problems, we have used the actual inside diameter, thereby introducing a small error. In most instances the error introduced is well within the limits of accuracy of the data. For certain cases (particularly small pipe sizes) this may not be true, and the data given in Appendix C for pipe sizes should be used.

The second caution concerns the use of the Bernoulli equation. Due to the nonuniform velocity distribution across any section of a pipe, a correction should be made in the velocity head terms. Again, in most cases of concern to engineers this correction can be neglected. However, in laminar flow and turbulent flow with a low Reynolds number this correction may be significant.

6.6 Hazen-Williams Formula

Prior to the works of Nikuradse, Pigott, Kemler, and Moody, many empirical equations were developed and used for solving problems involving the flow of water in pipes. The most widely used empirical formula is the Hazen–Williams formula:

$$V = 1.32CR^{0.63}S^{0.54} \qquad (6.37)$$

where C is the Hazen-Williams coefficient, R the hydraulic radius (area per perimeter), and S the friction loss per unit length of pipe. Some typical values for C are 140 for extremely straight and new pipes, 110 for new riveted steel, and 60–80 for old pipes in bad condition. This equation is still used in the field of water supply, and extensive tabulations and charts have been devised for its solution. While the principal advantage of this equation is that all problems can be solved directly, it has the severe disadvantage of being inapplicable to all fluids and its empirical nature. (The Hazen-Williams formula was devised from empirical test data obtained for water flowing in pipes. The formula does not have a fully rational basis and in principle its use should be restricted to water.) The student is admonished to use the more rational approach developed earlier in this chapter rather than the empirical Hazen-Williams formula.

ILLUSTRATIVE PROBLEM 6.9. Solve illustrative problem 6.5 using the Hazen-Williams formula and assuming $C = 140$.

Solution. The ratio of area to perimeter is half the radius of a circular pipe. Therefore,

$$V = 1.32(140)\left(\frac{2.067}{4 \times 12}\right)^{0.63}\left(\frac{50}{200}\right)^{0.54}$$
$$= 12.0 \text{ ft/sec}$$

The solution is readily arrived at; there is reasonably good agreement in the velocity calculated by the empirical Hazen-Williams formula (12.0 ft/sec) and the value obtained from the rational approach (11.8 ft/sec).

ILLUSTRATIVE PROBLEM 6.10. Solve illustrative problem 6.4 using the Hazen-Williams formula and assuming $C = 110$ for concrete.

Solution. From illustrative problem 6.4, $V = 6.3$ ft/sec. Therefore,

$$6.3 = 1.32(110)\left(\frac{18}{12 \times 2}\right)^{0.63} S^{0.54}$$

$$S^{0.54} = 0.052$$

$$S = 0.0042 \text{ ft head loss per foot of pipe}$$

and for 1000 ft of pipe, the frictional head loss is 4.2 ft of water. The agreement with illustrative problem 6.4 is reasonably good. For this problem the choice of C becomes critical; i.e., for C of 140, the solution is 2.7 ft of water, which is grossly in error.

6.7 Intersecting Pipes

When pipes are connected in parallel, as shown in Figure 6.19, we can state

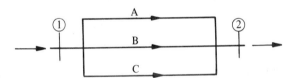

Figure 6.19. *Pipes in parallel.*

that the sum of the flows in the three branches must equal the total flow in the main, and loss in pressure between sections 1 and 2 must be the same regardless of the path taken. It is interesting to note the analogy that this problem has with a direct current network in which the circuit elements are resistors connected in parallel. We can compare the current in the electrical circuit with the quantity of fluid flowing in the pipes; the resistance (in ohms) of the electrical circuits has its counterpart in the flow resistance in the pipes, and the potential drop (voltage) in the electrical circuit is analogous to the pressure drop in the pipe. This type of electrical analogy has provided the basis for the solution of many fluid flow problems on analog computers.

If we consider any of the branches such as branch A between points 1 and 2 in Figure 6.19, we can write the following:

$$Q_A = A_A V_A \tag{6.38a}$$

and

$$V = K'_A \sqrt{h} \tag{6.38b}$$

where h is the loss of head between sections 1 and 2, K' is $\sqrt{2g/K_{eq}}$, and K_{eq} is the sum of all the loss coefficients in branch A. For each of the other branches we can write similar relations. Thus we can write for the flow in the main, Q,

$$Q = Q_A + Q_B + Q_C = K'_A\sqrt{h} + K'_B\sqrt{h} + K'_C\sqrt{h}$$

or

$$Q = (K'_A + K'_B + K'_C)\sqrt{h} \qquad (6.39)$$

Equation (6.39) can be used to obtain either the flow in the main or the flow in any of the branches.

ILLUSTRATIVE PROBLEM 6.11. A steel horizontal branch piping system is as shown in the accompanying diagram. Neglect minor losses and determine the pressure drop from 1 to 2 if water at 68°F is flowing into the system at section 1 at the rate of 750 gal/min.

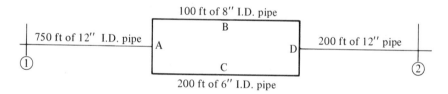

Solution. The pressure drop in branches A and D can be determined explicitly as follows: For branch A,

$$750 \text{ gal/min} \times \frac{231}{1728} = 100.5 \text{ ft}^3/\text{min}$$

$$\text{Flow area} = \frac{\pi}{4}(1)^2 = 0.7854 \text{ ft}^2$$

$$\text{Velocity} = Q/A = \frac{100.5}{60 \times 0.7854} = 2.13 \text{ ft/sec}$$

$$VD = 2.13 \times 12 = 25.5$$

$$e = 0.00015, \qquad e/D = 0.00015$$

From Figure 6.11,

$$\text{Re} = 1.7 \times 10^4, \qquad f = 0.027$$

$$\Delta p_f = f\frac{L}{D}\frac{V^2}{2g}\frac{\gamma}{144} = 0.027 \times \frac{750}{1} \times \frac{2.13^2}{2g} \times \frac{62.4}{144} = 0.617 \text{ psi}$$

For Branch D, all calculations are the same as for branch A with only the length changed:

$$\therefore \Delta p_f = 0.617 \times \frac{1200}{750} = 0.99 \text{ psi}$$

Branches B and C cannot be solved explicitly since the quantity flowing in each branch is unknown. However,

$$Q_B + Q_C = 750 \text{ gal/min}$$

Thus,

$$Q_B = 750 - Q_C \text{ gal/min}$$
$$Q_B = (750 - Q_C)\frac{231}{1728}\frac{1}{60} \text{ ft}^3/\text{sec}$$

and

$$Q_B = 1.67 - 0.00223 Q_C \text{ ft}^3/\text{sec}$$

However,

$$V_B = \frac{Q_B}{A_B} = \frac{Q_B}{\frac{\pi}{4}\left(\frac{8}{12}\right)^2} = 2.87 Q_B$$

Then

$$V_B = 4.79 - 0.0064 Q_C$$

We also have the condition that

$$(\Delta p_f)_B = (\Delta p_f)_C$$

As a first approximation let $f_B = f_C$. Then

$$\left(\frac{L}{D}\right)_B \frac{V_B^2}{2g} = \left(\frac{L}{D}\right)_C \frac{V_C^2}{2g}$$

Substituting for V_B and rearranging, we have the following quadratic equation for Q_C:

$$Q_C^2 + 204 Q_C - 76{,}000 = 0$$

and

$$Q_C \simeq 190 \text{ gal/min}$$

Therefore,

$$Q_B \simeq 560 \text{ gal/min}$$

Now check branch C:

$$V = \frac{190 \times 231}{1728 \times 60} \frac{1}{0.196} = 2.16 \text{ ft/sec}$$

$$VD = 6 \times 2.16 = 13$$

$$\frac{e}{D} = \frac{0.00015}{12} = 0.0003$$

$$\text{Re} = 9 \times 10^4$$

$$f = 0.02$$

$$(\Delta p_f)_C = 0.02 \times \frac{200}{\frac{1}{2}} \times \frac{(2.16)^2}{2g} = 0.58 \text{ ft}$$

Now check branch B:

$$V_B = \frac{560}{60} \times \frac{231}{1728} \frac{1}{\frac{\pi}{4}\left(\frac{8}{12}\right)^2} = 3.59 \text{ ft/sec}$$

$$VD = 3.59 \times 8 = 28.7$$

$$\frac{e}{D} = \frac{0.00015}{\frac{8}{12}} = 0.000225$$

$$\text{Re} = 2 \times 10^5$$

$$f = 0.0178$$

$$(\Delta p_f)_B = 0.0178 \times \frac{(3.59)^2}{2g} \times \frac{100}{\frac{8}{12}} = 0.53 \text{ ft}$$

At this point we note that the two pressure drops are not equal (as they must be), and if the desired accuracy warrants the effort, another iteration can be made using the results of the foregoing solution. The quantity of water flowing in branch C will be less than 190 gal/min, and in branch B it will be greater than 560 gal/min. The total pressure drop between 1 and 2 will be, closely,

$$0.617 + 0.99 + \frac{0.55(62.4)}{144} = 1.85 \text{ psi}$$

The preceding problem is a relatively simple problem, and it can be seen that a considerable expenditure of time and effort is required before a solution can be obtained. For more complex problems the time required to obtain a solution in the manner indicated is prohibitive. Numerical precedures have been devised to decrease the effort in problems of this type, but these procedures can become quite lengthy, too. Fortunately these problems can be programmed relatively easily for solution on both analog and digital computers, and at present almost all complex problems of this type are solved using computers. It is important that the student understand the basis of this class of problem so that the necessary inputs for the computerized solution can be obtained.

6.8 Closure

In this chapter we have placed the emphasis on the practical aspects of calculating the pressure drop in pipes and ducts when the fluid flowing is incompressible. While the approach to this problem has been on a rational basis, it has been necessary to invoke empirical correlations to permit the engineer to make quantitative evaluations in commercial piping systems. Even with the best available data it is possible to find deviations in excess of 30 percent between calculated and measured results for new or clean commercial pipes. For old or severely corroded pipes it is almost impossible to obtain a usable engineering correlation for pressure drop.

Pressure changes occur at changes in section, valves, bends, branches, fittings, inlets, and outlets and must be included with the friction pressure drops in any calculation of pump size or head requirements. For these losses (termed other or minor losses) the available correlations leave much to be desired, and in the worst instance a variation of $+80$ percent -200 percent can be found for a flanged check valve. It is evident that the designer of a piping system must exercise a great deal of judgment in evaluating the pressure drop and must allow for this lack of accurate data in sizing his pumps or establishing the head on a given system. The novice or student should be aware of the inadequacies of our knowledge in this branch of fluid mechanics and should not depend blindly on a calculated flow rate or pressure drop.

REFERENCES

1. *Pipe Friction Manual*, 3rd ed., Hydraulic Institute, New York 1961.
2. "Fluid Mechanics" by A. D. Kraus, *Electro-Technology* (New York), **69**, April 1962, p.120.

3. "Flow of Fluids Through Valves, Fittings, and Pipe," *Technical Paper #410*, Crane Co., Chicago, 1957, p. 179.

4. Flow Losses in Abrupt Enlargements and Contractions" by R. P. Benedict, N. A. Carlucci, and S. D. Swetz, *ASME Paper 65-WA/PTC-1.*

5. *Engineering Heat Transfer* by S. T. Hsu, Litton Educational Publishing, Inc., New York, 1963.

6. "Friction Factors for Pipe Flow" by L. F. Moody, *Transactions of the ASME*, **66**, 1944, p. 671; also *Mechanical Engineering*, **69**, December 1947, p. 1005.

7. "The Flow of Fluids in Closed Conduits" by R. J. S. Pigott, *Mechanical Engineering*, **55**, No. 8, August 1933, p. 497.

8. "A Study of Data on the Flow of Fluids in Pipes" by E. Kemler, *Transactions of the ASME*, **55**, 1933, HYD-55, p. 7.

9. "Laws of Flow in Rough Pipes" by J. Nikuradse, *NACA Technical Memorandum #1292*, November 1950.

10. *Engineering Applications of Fluid Mechanics* by J. C. Hunsaker and B. G. Rightmire, McGraw-Hill Book Company, Inc., New York, 1947.

11. *Themodynamics of Fluid Flow* by N. A. Hall, Prentice-Hall, Inc., Englewood Cliffs, N. J., 1951.

12. *Introduction to Gas Dynamics* by R. M. Rotty, John Wiley & Sons, Inc., New York, 1962.

13. *Fluid Mechanics* by R. C. Binder, 4th ed., Prentice-Hall, Inc., Englwood Cliffs, N. J., 1962.

14. *Fluid Mechanics* by V. L. Streeter, 3rd ed., McGraw-Hill Book Company, Inc., New York, 1962.

15. *Elementary Fluid Mechanics* by J. K. Vennard, John Wiley & Sons, Inc., New York, 1961.

16. *Mechanical Engineering* by G. Coren, Arco Publishing Co., Inc., New York, 1964.

PROBLEMS

6.1 Show that for a given volumetric flow that the pressure drop in a pipe is inversely proportional to the forth power of the diameter in laminar flow and inversely proportional to the fifth power of the diameter for turbulent flow.

6.2 Evaluate $r^3 \, \Delta r$ between the limits of $r = 4$ and $r = 1.5$ either numerically or graphically and compare the result with the exact value.

6.3 Derive an equation for the equivalent diameter of a circular pipe when the fluid partially fills the pipe. Assume that the pipe diameter is D and that the distance from the center of the pipe to the fluid level is h.

6.4 A sphere of graphite (soot) having a specific weight of 138 lb/ft^3 falls in air. If its diameter is 0.5 mm, with what velocity will it settle in still air at 68°F?

6.5 A spherical ball is dropped in oil. If the viscosity of the oil is 2 P, its specific weight is 58 lb/ft^3, and the diameter of the ball is 0.1 cm, determine the terminal velocity of the ball. $\gamma_S = 477$ lb/ft^3.

6.6 If Stokes' law is applicable up to a Reynolds number of approximately 2, determine the maximum diameter of a spherical carbon particle ($\gamma = 138$ lb/ft^3) that will settle in still air at 68°F and still conform to Stokes' law.

6.7 A fluid flows in a circular tube having an inside diameter of $\frac{1}{8}$ in. If the viscosity of the fluid is 5 cP and its average velocity is $\frac{1}{2}$ ft/sec, determine the shear stress at the tube wall.

6.8 Oil flows through a small size tube having a $\frac{1}{4}$-in. i.d. It is desired that the flow shall always be laminar. If the viscosity of the oil is 1 P and its weight is 52 lb/ft^3, determine the maximum velocity in the tube.

6.9 Assuming that the concept of an equivalent diameter can be applied to an annulus between two concentric tubes having diameters D_2 and D_1, determine the equivalent diameter of such an annulus in terms of D_2 and D_1, assuming that the fluid wets both surfaces.

6.10 Based upon the results of problem 6.9 ($D_{eq} = D_2 - D_1$), calculate the pressure drop per foot of an annulus having D_2 equal to 1 in. and D_1 equal to $\frac{3}{4}$ in. if the fluid flowing is an oil with a viscosity of 90 cP and a specific weight of 50 lb/ft^3. The average flow velocity is 10 ft/sec.

6.11 A fluid flows in a horizontal pipe at the rate of $\frac{1}{4}$ ft^3/sec, and the friction head loss is measured to be 25 ft. If the pipe is 1000 ft long and has an inside diameter of 2 in., determine whether the flow is laminar or turbulent.

6.12 Oil is being pumped from a truck to a tank 10 ft higher than the truck through a 2-in.-i.d. pipe 100 ft long. Assume that the pump outlet pressure is 15 psig and determine the flow rate in gallons per minute. The oil has an absolute viscosity of 105 cP and a specific weight of 56 lb/ft^3. (*Hint:* Assume the flow is laminar and check after solving the problem.)

6.13 A company wishes to pump crude oil from a storage tank to its plant 35 miles away. Assume that an 18-in.-i.d. pipeline is to be used and that the plant is at an elevation of 50 ft above the tank. The oil has a viscosity of 150 cP and a specific weight of 52 lb/ft^3. Determine the required pressure at the pump if the average velocity in the pipe is to be limited to 1.5 ft/sec.

6.14 Determine the pressure drop in a 2-in.-i.d. pipe that is 1000 ft long in which oil flows at an average velocity of 3 ft/sec if the specific weight of the oil is 51 lb/ft^3 and its viscosity is 56 cP.

6.15 Oil flows in a 3-in.-i.d. pipe. If the absolute viscosity of the oil is 92 cP and its specific weight is 50 lb/ft³, determine the flow rate, Reynolds number, and pressure drop per foot of length if the maximum velocity is 4 ft/sec.

6.16 A commercial steel pipe whose inside diameter is 6 in. is used to deliver water at the rate of 700 gal/min to a chemical plant. Determine the friction loss per mile of pipe. Assume water at 68°F.

6.17 In problem 6.16, determine the friction loss per mile if there are two regular flanged elbows, three wide open flanged gate valves, a flanged check valve, and four 90-deg flanged long radius elbows.

6.18 An 8-in.-i.d. pipe is 2500 ft long and carries water from a pump to a reservoir whose water surface is 200 ft above the pipe. If 4 ft³/sec is being pumped, determine the gage pressure at the discharge of the pump. Assume that the pipe is rough cast iron and neglect minor losses.

6.19 Water at 68°F flows in a horizontal steel pipe at the rate of 15 gal/min. If the pipe has an inside diameter of 3 in., is 500 ft long, has three 90-deg regular elbows, four partially open gate valves, three regular 180-deg long radius bends, and a check valve, determine the pressure drop in the line. All fittings are flanged.

6.20 Water at 68°F flows in a vertical square steel pipe whose sides are 4 in. Determine the difference in pressure between two static pressure taps located 100 ft apart if the average water velocity is 15 ft/sec and the flow is up.

6.21 Oil having a viscosity of 50 cP and a specific weight of 52 lb/ft³ flows upward in a 1-in.-i.d. pipe with a velocity of 0.1 ft/sec. Determine the difference in pressure between two static pressure taps located 100 ft apart.

6.22 A diffuser is placed in a pipeline to connect a 6-in.-i.d. pipe to a 12-in.-i.d. pipe. If the total included angle of the diffuser is 30 deg and the flow is from the 6-to the 12-in. pipe at 10 ft/sec, determine the pressure drop in the diffuser.

6.23 If the diffuser included angle is 7 deg in problem 6.22, determine the pressure drop in the diffuser.

6.24 In problem 6.20 the flow is reversed. Determine the pressure drop.

6.25 Air at 50 psia and 100°F flows in a horizontal 6-in.-i.d. pipe. It is desired to limit the pressure drop due to pressure losses to 10 percent of the initial absolute air pressure. What is the maximum velocity in the pipe if minor losses are neglected and the pipe is 500 ft long?

6.26 An abrupt contraction is placed in a horizontal pipeline that has an inside diameter of 4 in. If the contracted flow diameter is 2 in., determine the pressure loss for 68°F water flowing at 10 ft/sec in the 4-in. pipe.

6.27 In problem 6.26 the flow is reversed. Determine the pressure drop if the velocity in the 4-in. line is kept at 10 ft/sec.

6.28 Water is supplied from a reservoir to a 500-ft-long horizontal round concrete pipe 36 in. in diameter. What head is required to cause a flow of 20,000 gal/min? Neglect entrance and exit losses.

6.29 Compare the results of problem 6.28 with the result obtained if the entrance and exit loss had not been neglected.

6.30 If water at 68°F flows in a 2-in.-i.d. pipe that is 250 ft long, determine the flow in gallons per minute. Neglect minor losses.

6.31 Water at 68°F flows in a 2-in.-i.d. pipe at the rate of 250 gal/min. If the pipe has a globe valve and a check valve and it delivers water to a tank that is 60 ft above the pipe inlet, determine the pressure difference between the pipe inlet and outlet. All fittings are screwed; neglect inlet and outlet losses.

NOMENCLATURE

A = area
a = constant
C = Hazen-Williams coefficient
D = diameter
e = absolute roughnes
F = force
f = coefficient of friction
g = local acceleration of gravity
h = head loss
h_f = friction head loss
K = loss coefficient
K' = loss coefficient
L = length
P = perimeter
p = pressure
Q = volume rate of flow
R = hydraulic radius
Re = Reynolds number = $DV\rho/\mu$
r = tube radius
S = head loss per foot length of pipe

V = velocity
W = Weight
x = horizontal displacement
γ = specific weight
Δ = small increment
θ = angle
μ = coefficient of viscosity
ρ = density
τ = shear stress

Subscripts
   eq = equivalent
   F = fluid
   o = outer or wall
   s = solid
   1, 2, 3 = states or locations

Superscripts
   b, e, f, g, n = constants

Chapter **7**

INTRODUCTION TO COMPRESSIBLE FLOW[1]

7.1 Introduction

In recent years, with the advent of missiles, high-speed aircraft, and the flow of gases in such devices as gas turbines and rocket exhaust nozzles, the area of fluid mechanics denoted as compressible flow has taken on increased importance. To treat the general topic of compressible fluid flow in terms of modern developments would require volumes. However, many devices and flow regimes can be treated with reasonable approximation by considering the flow to be one-dimensional, that is, fluid properties are uniform over any cross section. This chapter considers the steady flow of gases based upon a one-dimensional model and the assumption that these gases conform to the ideal gas equation of state. Before proceeding, one additional observation must be made: The nomenclature in this chapter has been varied somewhat from the general nomenclature used

[1]Portions of this chapter are taken with permission from the author's text, *Elementary Applied Thermodynamics*, John Wiley & Sons, Inc., New York, 1965.

in the rest of the book to conform more nearly to the accepted nomenclature in the field. Symbols introduced in this chapter that differ from those used elsewhere in the text are defined as they are introduced.

7.2 Basic Principles

The study of one-dimensional steady compressible flow is based upon certain basic considerations. The velocity of a weak pressure wave propagated in the fluid is found to be important, and the student will note that the velocity of sound is but a familiar example of a moving weak pressure gradient in the fluid. The mode of propagation and the magnitude of the velocity of propagation of such a disturbance is at every point a function of the local properties of the fluid. The foregoing assumes that the pressure disturbance can be considered weak enough so that there is essentially no heat transfer and that any fluid friction arising from the velocity, temperature, or pressure gradients can be considered negligible. In essence, this is equivalent to assuming that this process is isentropic.

Before attempting to analyze the process of propagation of a weak pressure signal in a gas, let us attempt to establish the dependence of this event on the physical quantities involved using the principle of dimensional homogeneity as discussed in Chapter 5. We shall initially assume that at a given temperature the velocity of sound in a gas (a) is a function of the gas density (ρ), the gas pressure (p), and the viscosity (μ) of the gas. For this problem the gas viscosity has been introduced even though we doubt that it plays any role in the process under consideration. Proceeding as outlined in Chapter 5,

$$a = k\rho^b p^c \mu^d$$

where k is a dimensionless constant. Using the M-L-T-θ system, we have the dimensional equation

$$LT^{-1} = (ML^{-3})^b (ML^{-1}T^{-2})^c (MT^{-1}L^{-1})^d$$

Equating exponents,

1. M, $\quad 0 = b + c + d$.
2. L, $\quad 1 = -3b - c - d$.
3. T, $\quad -1 = -2c - d$.

Solving these equations yields $b = -\frac{1}{2}$, $c = \frac{1}{2}$, and $d = 0$. Since $d = 0$, we conclude that the velocity of sound in a gas is not dependent on the viscosity of the gas and arrive at the result that,

$$a = K\sqrt{\frac{p}{\rho}}$$

For an ideal gas $p/\rho = RT$ and

$$a = K\sqrt{RT} = k'\sqrt{T}$$

Therefore, for an ideal gas the velocity of sound (the velocity of propagation of a weak pressure signal) is a function only of the absolute temperature of the gas. Dimensional analysis does not yield the value of the dimensionless constant. It is therefore necessary to describe the process in an idealized manner and subsequently to mathematically analyze this idealization in order to develop fully an expression for the velocity of propagation of a weak pressure signal in a gas.

The following simplified physical model will help us to develop an expression for the velocity of propagation of a weak pressure signal in a gas. Assume, as is shown in Figure 7.1, that a piston at the end of a tube is moved an in-

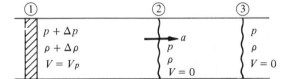

Figure 7.1. Piston position and disturbance at t = t.

finitesimal amount impulsively. This impulse will be transmitted through the fluid that occupies the duct. That portion of the fluid that has not been overtaken by the traveling disturbance will be at rest and have properties p and ρ. Behind the wave front, which is shown to have reached position 2 in Figure 7.1, the properties will vary from the undisturbed fluid by small amounts, that is, $p + \Delta p$ and $\rho + \Delta \rho$. The symbol Δ is used to denote a small quantity. The fluid at the piston face will move essentially with the velocity of the piston V_p. In the distance 1-2 the fluid will also have the disturbance velocity V_p. At another time greater than that shown in Figure 7.1 by the amount Δt the conditions shown in Figure 7.2 will be found. During this time the piston will have moved to a new position 1′, and the front of the disturbance will have moved to position 2′. Since the fluid ahead of the wave front is undisturbed, all of the

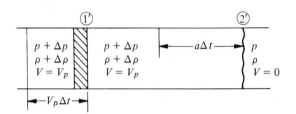

Figure 7.2. *Piston position and disturbance at* $t = t + \Delta t$.

mass originally contained between sections 1 and 3 must be the same as that contained between sections 1' and 2'. This is expressed mathematically as

$$A(\rho + \Delta\rho)V_p \,\Delta t = A(\Delta\rho)(a \,\Delta t) \tag{7.1}$$

where a is the velocity of propagation of the disturbance. Products of the Δ's in equation (7.1) will be discarded as being negligibly small. Simplifying,

$$a = V_p \frac{\rho}{\Delta\rho} \tag{7.2}$$

The relation between the piston velocity and the velocity of propagation of the disturbance can be obtained by applying Newton's second law of motion to the mass being accelerated, with the piston and the disturbance both moving with constant velocity. The force must be the unbalanced force on the mass in question. By noting that the pressure to the left of the disturbance is $p + \Delta p$ and to the right p, the unbalanced force F is

$$F = (p + \Delta p)A - pA = A \,\Delta p \tag{7.3}$$

and the increment of mass is

$$m = (\rho + \Delta\rho)A(a \,\Delta t) - (\rho + \Delta\rho)AV_p \,\Delta t \tag{7.4}$$

Therefore,

$$A \,\Delta p = \frac{\{(\rho + \Delta\rho)[Aa(\Delta\rho) - AV_p(\Delta t)]\}V_p}{(\Delta t)g} \tag{7.5}$$

By expanding (7.5) and neglecting powers of Δ the desired relation is obtained: that is,

$$V_p = g\frac{\Delta p}{\rho a} \tag{7.6}$$

By combining (7.2) with (7.6) and noting that the process was postulated to be isentropic, it is possible to write the more general equation as

$$a^2 = g\left(\frac{\Delta p}{\Delta \rho}\right)_s \tag{7.7}$$

Equation (7.7) is the general equation that will apply regardless of the nature of the gas, as long as it is homogeneous. For an ideal gas it is a simple matter to place this equation in a more desirable form by using the equation of state of the gas and the equation of the isentropic path as follows:

$$p = R\rho T \tag{7.8}$$

and

$$p = (\text{Constant})\rho^k \tag{7.9}$$

For the ideal gas, it can be shown that

$$\left(\frac{\Delta p}{\Delta \rho}\right)_s = \text{Constant } k(\rho^{k-1}) \tag{7.10}$$

In equation (7.10), k has been assumed to be constant. By noting that

$$\rho^{k-1} = \frac{\rho^k}{\rho} = \frac{p}{\text{Constant}}$$

equation (7.11) is achieved by replacing ρ^{k-1} with $p/\text{constant}$:

$$\left(\frac{\Delta p}{\Delta \rho}\right)_s = \frac{kp}{\rho} = kRT \tag{7.11}$$

The square of acoustic velocity or the square of velocity of propagation of a small disturbance in an ideal gas is therefore

$$a^2 = g\frac{kp}{\rho} = gkRT; \qquad a = K'\sqrt{T} \tag{7.12a}$$

For air

$$a = 49.1\sqrt{T} \qquad \text{ft/sec when } T \text{ is in } °R \tag{7.12b}$$

It will be noted that these results agree with the result obtained using dimensional analysis with $K' = \sqrt{gkR}$.

ILLUSTRATIVE PROBLEM 7.1. Determine the velocity of sound in air at 100, 500, and 1000°F. What is the velocity of sound at these temperatures if the gas is hydrogen?

Solution. For air, equation (7.12b) is applicable:

$$a = 49.1\sqrt{T}$$

$t(°F)$	T	$\sqrt{T}$	a_{air} (ft/sec)	$a_{hydrogen}$
100	560	23.7	1165	4430
500	960	31.0	1525	5810
1000	1460	38.2	1880	7170

For any other gas that can be treated as an ideal gas $K' \propto \sqrt{kR}$. R in engineering units is 1545/molecular weight, and k can be found in standard tables for various gases.[2] Using Table A-13, p. 271 of *Elementary Applied Thermodynamics* by Granet (see footnote 1), we have $R = 766.53$ and $k = 1.41$ for hydrogen, and for air we have $R = 53.35$ and $k = 1.40$. Therefore,

$$a_{hydrogen} = a_{air}\sqrt{\frac{766.53 \times 1.41}{53.36 \times 1.40}} = 3.8a_{air}$$

The corresponding values of $a_{hydrogen}$ for the temperatures of this problem are tabulated above.

In the subseqent work in this chapter it will be found that the Mach number is an important parameter; it is defined as the ratio of the velocity at a point in a fluid to the velocity of sound at that point at a given instant of time. Denoting the local fluid velocity by V and the velocity of sound as a,

$$M = \frac{V}{a} \tag{7.13a}$$

and

$$M^2 = \frac{V^2}{gk\left(\dfrac{p}{\rho}\right)} = \frac{V^2}{gkRT} \tag{7.13b}$$

where equation (7.13b) is applicable only to an ideal gas.

ILLUSTRATIVE PROBLEM 7.2. Air is flowing in a duct at atmospheric pressure with a velocity of 1500 ft/sec. If the air temperature is 200°F, what is the Mach number?

Solution. Using equation (7.12b),

[2]See Chapter 3.

$$a = 49.1\sqrt{T} = 49.1\sqrt{660} = 1262 \text{ ft/sec}$$

By definition,

$$M = \frac{V}{a} = \frac{1500}{1262} = 1.188$$

7.3 General Relations

The equation of continuity for the steady flow of a fluid states that mass rate of flow in an apparatus is a constant in both time and space. Mathematically,

$$\dot{m} = \rho AV = \text{Constant} \tag{7.14a}$$

Now permit p, A, and V to vary by small amounts even though $\dot{m}$ is constant. Therefore,

$$\dot{m} = (\rho + \Delta\rho)(A + \Delta A)(V + \Delta V) \tag{7.14b}$$

By subtracting (7.14a) from (7.14b) and dividing through by ρAV,

$$\frac{(\rho + \Delta\rho)(A + \Delta A)(V + \Delta V) - \rho AV}{\rho AV} = 0 \tag{7.15}$$

If the numerator of equation (7.15) is multiplied as indicated and the products of the Δ's are neglected as being very small,

$$\frac{\Delta\rho}{\rho} + \frac{\Delta A}{A} + \frac{\Delta V}{V} = 0 \tag{7.16}$$

Equation (7.16) expresses the fractional change in each variable as a function of the other variables in the continuity equation. This particular form is convenient for present purposes. For example, for a constant area duct $\Delta A = 0$, and equation (7.16) becomes

$$\frac{\Delta\rho}{\rho} + \frac{\Delta V}{V} = 0 \tag{7.17}$$

If the flow in a constant area duct is also isothermal, an interesting relation is obtained by noting from equation (7.13a) that $M^2 \propto V^2$ and $M\,\Delta M \propto V\,\Delta V$. Putting this into equation (7.17),

$$\frac{\Delta p}{\rho} + M\frac{\Delta M}{V^2} = 0 \quad \text{or} \quad \frac{\Delta p}{\rho} + \frac{\Delta M}{M} = 0 \tag{7.18}$$

This form of equation is similar to the form of equation (7.16). Thus for isothermal constant area flow,

$$\rho_1 M_1 = \rho_2 M_2 = \text{Constant} \tag{7.19a}$$

$$\rho_1 V_1 = \rho_2 V_2 = \text{Constant} \tag{7.19b}$$

Of course, by noting that this relation is for an ideal gas flowing isothermally it can be readily expressed in terms of pressure by using the equation of state for the ideal gas. Therefore,

$$\frac{\Delta p}{\rho} + \frac{\Delta M}{M} = 0 \tag{7.20}$$

In solving problems in fluid mechanics, the assumption is frequently made that a fluid flowing in a pipe adiabatically may be treated as an isothermal incompressible fluid if the ratio of the pressure loss to the inlet pressure is less than 10 percent. The use of this approximation under these conditions is justified from equations (7.14a) and (7.20), and for many cases of importance to the engineer it represents a useful and time-saving approximation. It is necessary in questionable cases to verify that the assumptions are met and whether they are applicable. Too often the approximation is used without justification and can lead to gross errors.

7.4 Adiabatic Flow

In addition to the continuity equation and the equation of state for the gas, it is also necessary to utilize the energy equation for steady flow. Consider the special case of adiabatic flow without shaft work or elevation change. For this process the first law of thermodynamics yields

$$h + \frac{V^2}{2gJ} = h_1 + \frac{V_1^2}{2gJ} = h^o \tag{7.21}$$

Equation (7.21) defines the term stagnation enthalpy, h^o. This terminology is best illustrated by referring to Figure 7.3. At section "o" it is assumed that the area can be considered infinite or that the velocity is essentially zero. Thus from equation (7.21) the stagnation enthalpy is $h + V^2/2gJ$; $h_1^o = h_1 + V_1^2/2gJ$ and

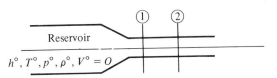

Figure 7.3

$h^o = h_1^o = h_2^o = $ constant. The stagnation enthalpy (or total enthalpy) is therefore a constant by definition for the process in question if it is adiabatic, if no work is done on or by the fluid, and if the fluid does not change in elevation.

The introduction of the terminology of the preceding paragraph requires a brief explanation. By referring to Figure 7.4 it will be noted that an isentropic

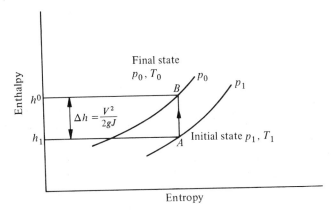

Figure 7.4

compression is shown on enthalpy-entropy coordinates by the line A, B. Since equation (7.21) is applicable, it follows that the change in enthalpy indicated in the diagram must equal $V^2/2gJ$, since the velocity at o conditions is zero.

The final state (the o state) is commonly known as the "total" state and the pressure, temperature, and density are known, respectively, as the total pressure, total temperature, and total density.

However, if the process is adiabatic but not isentropic, it may be shown as in Figure 7.5.

The final enthalpy will be the same as that for the isentropic case, but the entropy will be greater, and consequently the final pressure at the end of the actual process will be less. The end state of the nonisentropic (actual) process is commonly called the stagnation state and the properties at this state are called stagnation properties. By noting that the final enthalpy shown in Figures 7.4 and 7.5 are equal, either the terminology of *total* or *stagnation enthalpy* for this state is correct. Again, for the ideal gas certain relevant and important con-

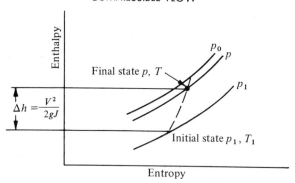

Figure 7.5

clusions may be drawn from this discussion. The final temperature of each process, that is, the total temperature and the stagnation temperatures, will be equal. The final pressures will not in general be equal. To recapitulate, the total state corresponds to an isentropic stagnation process and the stagnation state to an adiabatic process. Caution should be exercised in the use of this terminology. The literature contains many examples of differing definitions for these properties, and all terms used should be carefully defined.

ILLUSTRATIVE PROBLEM 7.3. Assume that a fluid is flowing in a device and that at some cross section the fluid velocity is 1000 ft/sec. If the fluid is saturated steam having an enthalpy of 1204.4 Btu/lb, determine its total enthalpy.

Solution. From the previous paragraph we can simply write

$$h^o - h = \frac{V^2}{2gJ}$$

For this problem

$$h^o - h = \frac{(1000)^2}{2 \times 32.2 \times 778} = 20 \text{ Btu/lb}$$

and

$$h^o = 1204.4 + 20 = 1224.4 \text{ Btu/lb}$$

It will be noted for this problem that if the initial velocity had been 100 ft/sec, Δh would have been 0.2 Btu/lb, and for most practical purposes the total properties and those of the flowing fluid would have been essentially the same. Thus for low-velocity fluids the difference in total and stream properties can be neglectected. At low velocities the fluid may be treated as though it were incompressible.

7.5 Isentropic Flow

Returning to equation (7.21) and permitting V to vary by an amount ΔV and h to vary by an amount Δh, we can readily show (problem 7.7) that

$$V\frac{\Delta V}{gJ} + \Delta h = 0 \tag{7.22}$$

For isentropic flow with no energy interchange as work and for no change in elevation,

$$\Delta h = \frac{\Delta p}{Jp}$$

since

$$\Delta h = C_p \Delta T = \frac{v\,\Delta p}{J} = \frac{\Delta p}{Jp} \tag{7.23}$$

By combining (7.22) and (7.23),

$$\frac{1}{\rho} = \frac{-V}{g}\left(\frac{\Delta V}{\Delta p}\right)_s, \tag{7.24}$$

where the subscript s denotes the fact that the process is isentropic.

At this point it is necessary to invoke the continuity relation given by equation (7.16). Inserting (7.24) in (7.16) and rearranging yields,

$$\frac{\Delta V}{V} = \frac{\dfrac{-\Delta A}{A}}{\left[1 - \dfrac{V^2}{g(\Delta p/\Delta\rho)_s}\right]} \tag{7.25}$$

By noting that the denominator contains $(V/a)^2$ and using the definition of Mach number from (7.13),

$$\frac{\Delta V}{V} = \frac{\dfrac{\Delta A}{A}}{(1 - M^2)} \tag{7.26}$$

By use of (7.16) the density relation becomes

$$\frac{\Delta\rho}{\rho} = -M^2\frac{\Delta V}{V} \tag{7.27}$$

Similarly,

$$\Delta p = Jp\,\Delta h = \frac{\dfrac{pV^2}{g}\dfrac{\Delta A}{A}}{(1 - M^2)} \qquad (7.28)$$

From equations (7.26) through (7.28) certain general conclusions can be reached for the isentropic flow of gas.

1. The relative change of density $\Delta\rho/\rho$ is M^2 times faster than the corresponding relative change in velocity $\Delta V/V$ and in the opposite direction. At Mach numbers greater than unity there are large density changes for small Mach number variations.
2. The variation in pressure is in the opposite direction of the relative change in velocity and is pV^2/g times greater.
3. When $M = 1$, $\Delta A = 0$. For small Mach numbers (less than unity) the relative change in velocity has the opposite sign compared to area changes. For Mach numbers always exceeding unity the velocity increases as the area increases.

7.6 Isentropic Flow of an Ideal Gas

For the isentropic flow of an ideal gas it is necessary only to recall that *ideal* refers to the equation of state of the gas and *isentropic* to the equation of the path. Equations (7.8) and (7.9), respectively, give the equation of state and path for an ideal gas flowing isentropically. By expanding the path equation for small change in p and ρ the following equation is derived:

$$\frac{\Delta p}{p} = k\frac{\Delta\rho}{\rho} \qquad (7.29)$$

Substituting (7.29) into (7.27) yields

$$\frac{\Delta p}{p} = \frac{k\,\Delta\rho}{\rho} = \frac{kM^2}{1 - M^2}\frac{\Delta A}{A} \qquad (7.30)$$

It is apparent from equations (7.29) and (7.30) that the relative change in p is k times the relative change in ρ.

We recall that for an ideal gas with constant specific heat it is possible to write the enthalpy (measured from a base of absolute zero) as

$$h^o = c_p T^o \qquad (7.31)$$

Equation (7.21) can be written in the following form:

$$V^2 = 2gJ(h^o - h_1) = 2gJc_p(T^o_1 - T_1) \qquad (7.32)$$

By applying the equations of path and state to (7.32) and simplifying,

$$V_2 = \left[2gJc_pT_1^o\left(1 - \frac{p_2}{p_o}\right)^{(k-1)/k}\right]^{1/2} = \left[\frac{2gk}{k-1}\frac{p^o}{p^o}\left(1 - \frac{p_2}{p_o}\right)^{(k-1)/k}\right]^{1/2}$$

$$(7.33)$$

Converging Nozzle. We apply equation (7.33) to the throat of the section of minimum area tt (Figure 7.6) and let p_t be the pressure at that section. The mass of gas passing through the section tt per unit time is

$$m = p_t A_t V_t = A_t p^o\left(\frac{p_t}{p^o}\right)^{1/k} V_t$$

$$= A_t\left\{\frac{2gk}{k-1}p^o\rho^o\left[\left(\frac{p}{p^o}\right)^{2/k} - \left(\frac{p}{\rho^o}\right)^{(k+1)/k}\right]\right\}^{1/2} \qquad (7.34)$$

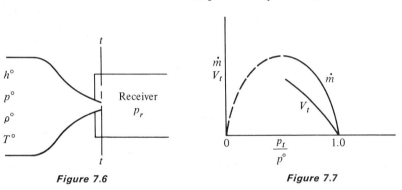

Figure 7.6   Figure 7.7

The variation of $\dot{m}$ with p/p^o is given by the curved line (partly dotted and partly solid) in Figure 7.7, which shows that $\dot{m}$ reaches a maximum value for a certain pressure ratio $p_t/p^o = p_c/p^o$; p_c is called the critical pressure and can be determined as follows: $\dot{m}$ is at its maximum when

$$\left[\left(\frac{p_t}{p_o}\right)^{2/k} - \left(\frac{p_t}{p^o}\right)^{(k+1)/k}\right]$$

is at a maximum.

By performing the required operations, using the methods of calculus, it is found that

$$\frac{p_t}{p^o} = \frac{p_c}{p^o} = \left(\frac{2}{k+1}\right)^{k/(k-1)} \qquad (7.35)$$

For air with $k = 1.4$, $p_c/p^o = 0.53$. For superheated steam with $k = 1.3$, $p_t/p^o = 0.546 = p_c/p$.

At the critical pressure ratio,

$$\frac{T_t}{T_0} = \frac{2}{k+1} \tag{7.36}$$

$$\frac{a_t}{a_o} = \sqrt{\frac{2}{k+1}} \tag{7.37}$$

$$V_t = \sqrt{\frac{2gk}{k+1}(p^o/\rho^o)} = \left(\sqrt{\frac{2}{k+1}}\right)(a_o) \equiv a_t \tag{7.38}$$

$$\dot{m}_{\max} = A_t\left(\frac{2}{k+1}\right)^{1/(k-1)}\sqrt{\frac{2gk}{k+1}(p^o/\rho^o)} \tag{7.39}$$

In general, also,

$$\left(\frac{A}{A_t}\right)^2 = \frac{k-1}{2}\left\{\frac{\left[\frac{2}{(k+1)}\right]^{(k+1)/(k-1)}}{\left(\frac{p}{p^o}\right)^{2/k}\left[1 - \left(\frac{p}{p^o}\right)^{(k-1)/k}\right]}\right\} \tag{7.40}$$

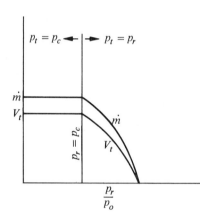

Figure 7.8

From the foregoing we can see that for a given nozzle throat area there is a maximum mass rate of flow of fluid that can pass through the nozzle. When the pressure ratio p_r/p^o is above the critical value, $\dot{m}$ increases with decreasing p_r. After p_r reaches p_c, there is no further increase in $\dot{m}$, and the flow is said to be choked. Under such conditions the fluid leaving the nozzle decreases in pressure from p_c to p_r through irreversible flow processes. These flow conditions are shown in Figure 7.8.

Converging and Diverging Nozzles (De Laval Nozzle). In the preceding discussion it was shown for a converging nozzle that the maximum velocity the fluid will attain at the exit section is the local sonic velocity and that the minimum pressure corresponding to sonic velocity at the exit is the critical pressure. To obtain a higher velocity and lower pressure at the exit section of a nozzle, a divergent portion is added downstream of the throat section. The fluid continues to expand in the divergent portion, reaching V_{e_1} and p_{e_1} at the exit. The fluid could also be compressed in the divergent portion, reaching V_{e_2} and p_{e_2} at the exit. The process in the nozzle is determined by the pressure in the receiver. If $p_r = p_{e_1}$, the former occurs. If $p_r = p_{e_2}$, the latter occurs. If p_r lies between the two, the fluid would first follow the former and reach p_r at the nozzle exit with a normal shock somewhere in the divergent portion of the nozzle or with compression waves in the receiver (see Figure 7.9).

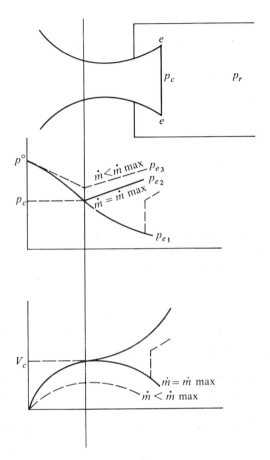

Figure 7.9

The foregoing discussion is for a mass flow rate in the nozzle equal to the maximum value. If the weight flow is below this value, the velocity at the throat is subsonic, and at the exit section of the nozzle $p = p_{e_s}$ and $V = V_{e_s}$.

To facilitate the use of many of the equations already developed, they have been placed into nondimensional form in terms of only two variables, namely the Mach number and the ratio of specific heats. Tables 30 and 53 of *Gas Tables* (reference 2) list the results of numerical computations of these one-dimensional compressible flow functions. In these tables the superscript $*$ refers to conditions in which $M = 1$ and the subscript o refers to isentropic stagnation. With this notation in mind, the working nondimensional equations are

$$T_o = \text{Constant} = T_o^* \tag{7.41}$$

$$T^* = \text{Constant} \tag{7.42}$$

$$p^o = \text{Constant} = p_o^* \tag{7.43}$$

$$p^* = \text{Constant} \tag{7.44}$$

$$M^* = \frac{V}{V^*} = M\left\{\frac{k+1}{2[1 + \frac{1}{2}(k-1)M^2]}\right\}^{1/2} \tag{7.45}$$

$$\frac{A}{A^*} = \frac{1}{M}\left\{\frac{2[1 + \frac{1}{2}(k-1)M^2]^{(k+1)/2(k-1)}}{k+1}\right\} \tag{7.46}$$

$$\frac{T}{T^*} = \frac{k+1}{2[1 + \frac{1}{2}(k-1)M^2]} \tag{7.47}$$

$$\frac{\rho}{\rho^*} = \left\{\frac{k+1}{2[1 + \frac{1}{2}(k-1)M^2]}\right\}^{1/(k-1)} \tag{7.48}$$

$$\frac{p}{p^*} = \left\{\frac{k+1}{2[1 + \frac{1}{2}(k-1)M^2]}\right\}^{k/(k-1)} \tag{7.49}$$

Table 7.1 is abridged from Table 30 of the Gas Tables and all of the foregoing nondimensional equations are tabulated as functions of Mach number. The convenience of these equations as well as their use is illustrated by illustrative problems 7.4 and 7.5. The application of Table 7.1 is also demonstrated in these problems.

ILLUSTRATIVE PROBLEM 7.4. An isentropic convergent nozzle is used to evacuate air from a test cell that is maintained at stagnation pressure and temperature of 300 psia and 800°R, respectively. The nozzle has inlet and outlet areas of 2.035 and 1 ft², respectively. Constant pressure specific heat is 0.24, the specific heat ratio is 1.4, and the gas constant R in consistent units is 53.35. Calculate the pressure, temperature, velocity, Mach number, and weight flow at the inlet and pressure and velocity at the outlet if the Mach number at the outlet is unity.

Solution. Refer to Figure 7.10. For $k = 1.4$ and $M = 1$ (Table 7.1),

$$\frac{T^*}{T_o} = 0.8333$$

$$\therefore T^* = 800(0.8333) = 686.6°R$$

and

$$a^* = \sqrt{gkRT^*} = \sqrt{32.2 \times 1.4 \times 53.35 \times 686.6}$$
$$= 1270 \text{ ft/sec} = V_2$$

Table 7.1 One-Dimensional Isentropic Compressible Flow Functions for an Ideal Gas with Constant Specific Heat and Molecular Weight and $k = 1.4$[†]

M	M^*	$\dfrac{A}{A^*}$	$\dfrac{P}{P_o}$	$\dfrac{\rho}{\rho^o}$	$\dfrac{T}{T_o}$
0	0	∞	1.00000	1.00000	1.00000
0.10	0.10943	5.8218	0.99303	0.99502	0.99800
0.20	0.21822	2.9635	0.97250	0.98027	0.99206
0.30	0.32572	2.0351	0.93947	0.95638	0.98232
0.40	0.43133	1.5901	0.89562	0.92428	0.96899
0.50	0.53452	1.3398	0.84302	0.88517	0.95238
0.60	0.63480	1.1882	0.78400	0.84045	0.93284
0.70	0.73179	1.09437	0.72092	0.79158	0.91075
0.80	0.82514	1.03823	0.65602	0.74000	0.88652
0.90	0.91460	1.00886	0.59126	0.68704	0.86058
1.00	1.00000	1.00000	0.52828	0.63394	0.83333
1.10	1.08124	1.00793	0.46835	0.58169	0.80515
1.20	1.1583	1.03044	0.41238	0.53114	0.77640
1.30	1.2311	1.06631	0.36092	0.48291	0.74738
1.40	1.2999	1.1149	0.31424	0.43742	0.71839
1.50	1.3646	1.1762	0.27240	0.39498	0.68965
1.60	1.4254	1.2502	0.23527	0.35573	0.66138
1.70	1.4825	1.3376	0.20259	0.31969	0.63372
1.80	1.5360	1.4390	0.17404	0.28682	0.60680
1.90	1.5861	1.5552	0.14924	0.25699	0.58072
2.00	1.6330	1.6875	0.12780	0.23005	0.55556
2.10	1.6769	1.8369	0.10935	0.20580	0.53135
2.20	1.7179	2.0050	0.09352	0.18405	0.50813
2.30	1.7563	2.1931	0.07997	0.16458	0.48591
2.40	1.7922	2.4031	0.06840	0.14720	0.46468
2.50	1.8258	2.6367	0.05853	0.13169	0.44444
2.60	1.8572	2.8960	0.05012	0.11787	0.42517
2.70	1.8865	3.1830	0.04295	0.10557	0.40684
2.80	1.9140	3.5001	0.03685	0.09462	0.38941
2.90	1.9398	3.8498	0.03165	0.08489	0.37286
3.00	1.9640	4.2346	0.02722	0.07623	0.35714
3.50	2.0642	6.7896	0.01311	0.04523	0.28986
4.00	2.1381	10.719	0.00658	0.02766	0.23810
4.50	2.1936	16.562	0.00346	0.01745	0.19802
5.00	2.2361	25.000	$189(10)^{-5}$	0.01134	0.16667
6.00	2.2953	53.180	$633(10)^{-6}$	0.00519	0.12195
7.00	2.3333	104.143	$242(10)^{-6}$	0.00261	0.09259
8.00	2.3591	190.109	$102(10)^{-6}$	0.00141	0.07246
9.00	2.3772	327.189	$474(10)^{-7}$	0.000815	0.05814
10.00	2.3904	535.938	$236(10)^{-7}$	0.000495	0.04762
∞	2.4495	∞	0	0	0

[†]Abridged from Table 30 in *Gas Tables* by Joseph H. Keenan and Joseph Kaye. Copyright 1948, by Joseph H. Keenan and Joseph Kaye. Published by John Wiley & Sons, Inc., New York.

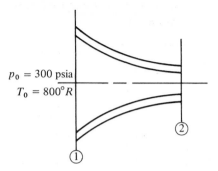

$M = 1$, therefore state ② is the * state   **Figure 7.10.**

For

$$\frac{A}{A^*} = 2.035$$

the tables yield $M_1 = 0.3$ and

$$\frac{p^*}{p_o} = 0.52828; \qquad \therefore p^* = 300(0.52828) = 158.5 \text{ psia}$$

Also,

$$\frac{T_1}{T_o} = 0.98232 \qquad \text{and} \qquad \frac{p_1}{p_o} = 0.93947$$

$$\therefore T_1 = 800(0.98232) = 785.9\,°R$$

and

$$p_1 = 300(0.93947) = 281.8 \text{ psia}$$

From the inlet conditions derived,

$$a_1 = \sqrt{gkRT_1} = \sqrt{32.2 \times 1.4 \times 53.35 \times 785.9} = 1375 \text{ ft/sec}$$

$$\therefore V_1 = M_1 a_1 = 0.3(1375) = 412 \text{ ft/sec}$$

The specific volume at inlet is found from the equation of state for an ideal gas:

$$v = \frac{RT_1}{p_1} = \frac{53.35 \times 785.9}{281.8 \times 144} = 1.035 \text{ ft}^3/\text{lb}$$

and

$$\dot{w} = \gamma A V = \frac{1}{1.035} \times 2.035 \times 412 = 810 \text{ lb/sec}$$

ILLUSTRATIVE PROBLEM 7.5. At a certain section of an air stream the Mach number is 2.5, the stagnation temperature is 560°R, and the static pressure is 0.5 atm. Assuming that the flow is steady isentropic and follows one-dimensional theory, calculate the following at the point at which M is 2.5: (1) temperature, (2) stagnation pressure, (3) velocity, (4) specific volume, and (5) weight velocity.

Solution. This problem will be solved by two methods (A and B).
A. *By equations.* Assume that $k = 1.4$ and $R = 53.3$:

$$a = \sqrt{gkRT} = \sqrt{32.2 \times 1.4 \times 53.3 \times 560} = 49.1\sqrt{560} = 775 \text{ ft/sec}$$
$$V = Ma = 2.5 \times 775 = 1935 \text{ ft/sec}$$

1. From equations (7.46) and (7.47),

$$\frac{T}{T^*} = \frac{k+1}{2[1 + \frac{1}{2}(k-1)M^2]}$$

but from (7.36),

$$\frac{T^*}{T_0} = \frac{2}{k+1}$$

Therefore,

$$\frac{T^*}{T_o} = \frac{1}{[1 + \frac{1}{2}(k-1)M^2]}$$

and

$$T = \frac{560}{1 + \frac{1}{2}(1.4-1)(2.5)^2} = \frac{560}{2.25} = 249°\text{R}$$

2.

$$\frac{p_o}{p} = \left(\frac{T_o}{T}\right)^{k/(k-1)}; \quad p_o = 0.5(14.7)\left(\frac{560}{249}\right)^{1.4/(1.4-1)} = 125.5 \text{ psia}$$

3. 1935 ft/sec.
4. From the equation of state,

$$v = \frac{RT}{p} = \frac{53.3 \times 249}{0.5(14.7)(144)} = 12.55 \text{ ft/}^3\text{lb}$$

5. Weight velocity is defined as the weight flow per unit area:

$$\frac{\dot{w}}{A} = \frac{\gamma A V}{A} = \gamma V = \frac{1935}{12.55} = 154.1 \ \text{lb/ft}^2/\text{sec}$$

B. *By Gas Tables.* At $M = 2.5$, Table 7.1 gives,

1.

$$\frac{T}{T_o} = 0.44444$$

2.

$$\frac{p}{p_o} = 0.05853$$

$$\therefore T = 560(0.44444) = 249°\text{R}$$

$$p = \frac{0.5 \times 14.7}{0.05853} = 125.5 \ \text{psia}$$

3. As before, 1935 ft/sec
4. As before, 12.55 ft³/lb.
5. As before, 154.1 lb/ft²/sec.

7.7 Actual Nozzle Performance

In the preceding sections certain idealizations were made regarding the character of the flow. Real fluids flowing in real devices deviate from these ideal conditions to some extent. There will always be frictional effects at the interface between the fluid and its containing walls; real fluids have viscosity and irreversible internal effects such as turbulence; temperature differences between the fluid and the walls give rise to nonadiabatic conditions; irreversible discontinuities (i.e., shocks) produce entropy increases, etc. Although it is beyond the scope of this book to investigate each of these irreversible processes, certain overall performance parameters have been established and are commonly used to express the performace of real fluids flowing in real nozzles.

The first of these performance indices is the nozzle efficiency. This parameter may be defined as

$$\eta = \frac{\text{Actual change in enthalpy}}{\text{Isentropic change in enthalpy}} \tag{7.50a}$$

or

$$\eta = \frac{\text{Actual kinetic energy at nozzle exit}}{\text{Kinetic energy at exit for an isentropic expansion}} \tag{7.50b}$$

for an expansion to the same pressure at the exit of the nozzle. It is important to note that equations (7.50a) and (7.50b) are strictly equivalent if the velocity entering the nozzle is zero or negligible. Figure 7.11 shows an isentropic ex-

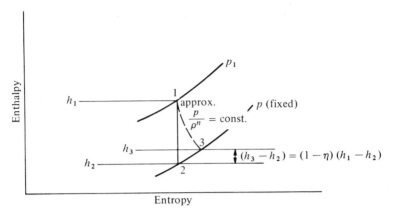

Figure 7.11.

pansion from state 1 to state 2. Also shown is a polytropic expansion characterized by a path equation $p/\rho^n = $ constant, where n is a function of the path. This approximation can be justified on the basis that the flow of a real fluid through a nozzle proceeds in such a rapid manner that it can be essentially adiabatic. Referring to Figure 7.11, we note that the nozzle efficiency (for negligible entrance velocity) is

$$\eta = \frac{h_1 - h_3}{h_1 - h_2} \quad \text{or} \quad \frac{T_1 - T_3}{T_1 - T_2} \tag{7.51}$$

for an ideal gas with constant specific heat. Equation (7.50a) can also be written in terms of stagnation enthalpy by use of equation (7.21)

$$\eta = \frac{h_1^o - h_3}{h_1^o - h_2}$$

$$= \frac{h_1 + \dfrac{V_1^2}{2gJ} - h_3}{h_1 + \dfrac{V_1^2}{2gJ} - h_2} \tag{7.50c}$$

When $V_1^2/2gJ$ is small, equations (7.50c) and (7.51) are essentially the same. Equation (7.50c) should be used when this velocity term is not small and cannot be considered negligible compared to the other terms in these equations.

For most nozzles the efficiency varies from 90 percent upward, with larger

nozzles having higher efficiencies. Fluids such as steam can show marked variations in efficiency, depending on the pressure and temperature range of operation.
Saturated steam will immediately expand into the wet region and a two-phase
fluid composed of steam and water will result. Superheated steam will behave
more nearly as an ideal gas until saturation is approached. At this time it will
commence to behave as saturated steam and ultimately go into two-phase flow
of steam and water. Because of this two-phase flow, it is possible for a real
nozzle to have a mass flow rate exceeding that for isentropic flow between the
same pressure ranges. This leads to the second performance index, namely, the
coefficient of discharge:

$$C_D = \frac{\text{Actual mass rate of flow}}{\text{Mass rate of flow for an isentropic expansion}} \tag{7.52}$$

As noted, C_D can exceed unity, but it is usually of the order of 95 percent.

The foregoing parameters do not completely specify the performance of real
fluids in real nozzles. This is readily shown from equation (7.52):

$$C_D = \frac{(\rho A V)_{\text{actual}}}{(\rho A V)_{\text{isentropic}}} = \frac{(\rho V)_{\text{actual}}}{(\rho V)_{\text{isentropic}}} \tag{7.53}$$

Turbine designers are interested in the exit velocity of the nozzle as well as its
mass discharge. Taking the square root of both sides of equation (7.51) yields

$$\sqrt{\eta} = C_V = \frac{\sqrt{h_1 - h_3}}{\sqrt{h_1 - h_2}} = \frac{\text{Actual exit velocity}}{\text{Isentropic exit velocity}} \tag{7.54}$$

for the same exit pressure. C_V is called the velocity coefficient. Equations (7.50),
(7.52), and (7.54), defining the efficiency, coefficient of discharge, and coefficient
of velocity, thus completely specify the exit conditions in the real nozzle. The
calculations of these coefficients is complex and beyond the scope of this study.
For further discussions of these coefficients and tabulated values, the interested
student is referred to the references at the end of this chapter.

7.8 Closure

In this chapter we have studied a flow regime in which the properties of
the fluid undergo large changes. To analyze this type of flow we have utilized
the continuity equation, the energy equation, and the equation of state for the
fluid, and to simplify the analysis we restricted our study to the one-dimensional

steady flow of an ideal gas. It was found that the Mach number is a nondimensional ratio that can be used to characterize the flow regime and to evaluate the fluid properties at any position in the flowing fluid. Fortunately, the one-dimensional steady flow of an ideal gas can be used to study many actual problems in the compressible flow regime. However, it is important to note that the results obtained are applicable only to the steady flow of a compressible fluid and should not be extended to the area of nonsteady (transient) flow.

The study of the flow of a compressible fluid required us to utilize thermodynamics and fluid mechanics in an integrated manner, and the student will note the close relationship between these two studies. At times it becomes totally artificial to draw a line of demarcation between thermodynamics and fluid mechanics; each is an integral part of the other, and this will be found to be the case if the references at the end of this chapter are consulted for further study of compressible flow.

REFERENCES

1. *Introduction to Aerodynamics of a Compressible Fluid* by H. W. Liepmann and A. E. Puckett, John Wiley & Sons, Inc., New York, 1947.

2. *Gas Tables* by J. H. Keenan and J. Kaye, John Wiley & Sons, Inc., New York, 1948.

3. *Thermodynamics of Fluid Flow* by N. A. Hall, Prentice-Hall, Inc., Englewood Cliffs, N. J., 1951.

4. *The Dynamics and Thermodynamics of Fluid Flow* by A. H. Shapiro, Vol. 1, The Ronald Press Company, New York, 1953.

5. *Elements of Gasdynamics* by H. W. Liepmann and A. Roshko, John Wiley & Sons, Inc., New York, 1957.

6. *Thermodynamics* by G. J. Van Wylen, John Wiley & Sons, Inc., New York, 1960.

7. *Jet Propulsion and Gas Turbines* by M. J. Zuchrow, John Wiley & Sons, Inc., New York, 1948.

8. *Principles of Engineering Thermodynamics* by P. J. Kiefer, G. F. Kenney, and M. C. Stuart, John Wiley & Sons, Inc., New York, 1954.

9. *Introduction to Gas Dynamics* by R. M. Rotty, John Wiley & Sons, Inc., New York, 1962.

10. *Engineering Mechanics* by F. L. Singer, 3rd ed., Harper & Row, Publishers, New York, 1954.

11. *Engineering Applications of Fluid Mechanics* by J. C. Hunsaker and B. G. Rightmire, McGraw-Hill Book Company, Inc., New York, 1947.

12. *Class Notes on Mechanics and Thermodynamics of Compressible Flow* by C. H. Wu, Polytechnic Institute of Brooklyn, Brooklyn, 1952.
13. *Mechanics and Properties of Matter* by R. J. Stephenson, John Wiley & Sons., Inc., New York, 1952.

PROBLEMS

7.1 Air at 50°F and 500 psia is flowing in a duct at 100 ft/sec. What is its Mach number?

7.2 A gas is flowing in a pipe at 200 ft/sec. If $C_p = 0.24$, determine the total and stagnation enthalpies and the total and stagnation temperatures if the temperature of the gas is 500°F. Use 0°F as a base for enthalpy.

7.3 Air is flowing at 500 ft/sec at a temperature of 300°F and a pressure of 25 psia. Determine its isentropic stagnation temperature and pressure.

7.4 Air is flowing in a duct at a Mach number of 2.5. Assuming that all processes are isentropic and the stagnation temperature is 560 R, determine the air temperature for which the Mach number is 1.2.

7.5 Air flows in a nozzle. If the inlet velocity is negligible and the inlet pressure is 100 psia, determine the critical velocity in the nozzle if it is a converging-diverging nozzle. Assume that the inlet temperature is 100°F. What is the critical pressure?

7.6 Air enters a nozzle at 150 psia and with negligible velocity. If the nozzle is frictionless, what is the Mach number at a section in which the pressure is 60 psia? Assume that $k = 1.4$

7.7 Air is slowed adiabatically from a velocity of 500 ft/sec to another velocity. During the slowing-down process the air temperature rises 10°F. If C_p of air is 0.24 BTU/lb °F and constant, determine its final velocity.

7.8 If air is slowed adiabatically from 500 to 300 ft/sec, determine its temperature rise if $C_p = 0.24$ BTU/lb °F.

7.9 Atmospheric air at 14.7 psia and 60°F leaks into a vacuum chamber through a partially open valve. If the valve opening is equivalent to a 1-in.-diameter hole, at what chamber pressure will sonic velocity occur in the valve? What weight rate of air does this correspond to?

7.10 Air is expanded from 150 psia and 600°F to 30 psia. If the exit Mach number is 1.7, what is the nozzle efficiency?

7.11 Derive equation (7.22).

7.12 Derive equation (7.23)

7.13 Problems 7.1 and 7.3 through 7.10 can be assigned using other gases such as nitrogen, hydrogen, helium, carbon dioxide, etc., and the results compared.

NOMENCLATURE

A = area

a = velocity of sound

C_D = coefficient of discharge

C_p = specific heat at constant pressure

C_v = coefficient of velocity

F = force

g = local acceleration of gravity

h = enthalpy

J = mechanical equivalent of heat

K = constant

k = isentropic exponent

M = Mach number

$\dot{m}$ = mass rate of flow

p = pressure

R = universal gas constant

s = entropy

T = absolute temperature

t = time

V = velocity

$\dot{w}$ = weight rate of flow

γ = specific weight

Δ = small increment

η = nozzle efficiency

μ = viscosity

ρ = density

Subscripts

c = critical

e = exit

o = stagnation

p = piston

r = receiver

s = entropy

t = throat

$1, 2, 3$ = states

Superscripts

k = isentropic exponent

o = stagnation state

$*$ = Mach 1

Chapter 8

OPEN CHANNEL FLOW

8.1 Introduction

The study of the fluid mechanics of rivers, culverts, canals, and conduits flowing partially full comes under the general heading of open channel flow. In this type of flow one surface is free and is not confined or in contact with a wall. Since the flow channel may be irregular and have a varying cross section and depth, the character of the flow differs from one location to another. The general considerations developed for the flow of fluids in pipes are applicable to open channel flow, but the presence of the free surface and the varying conditions in the channel make the analytical determination of the flow in open channels very difficult.

To approach this subject in a logical manner let us first identify and categorize the many possible flow regimes in a channel. The flow may be classified as being steady (independent of time) or unsteady (time dependent) and uniform or nonuniform; in addition, these flows can be further classified as tranquil or rapid. Steady uniform flow occurs in channels that are very long and whose depth and slope are constant along the length of the channel. In this type of flow, the slope of the free surface is found to be parallel to the slope of the bed of the channel. The flow in this case is also said to be tranquil if the velocity is low enough for a small wave to be able to travel upstream causing upstream conditions to be governed by downstream conditions; the flow is said to be rapid (or shooting) if the stream velocity is so high that a small wave cannot travel upstream. If the flow is such that the stream velocity is just equal to the velocity of a small wave, it is said to be critical.

Steady nonuniform flow occurs where the channel or its depth (or both) change from section to section. In this case the velocity must change from section to section, and the flow can change from tranquil to rapid or from rapid to tranquil at different cross sections along the length of the channel.

8.2 Uniform Steady Flow

Where fluid is flowing uniformly and steadily in an open channel, the rate of flow of fluid past any cross section as well as the cross section will be constant at all times and locations. Thus every section of the channel will appear to be the same as every other section. The energy that provides the driving force for the flow is the change in potential energy of the channel as it slopes

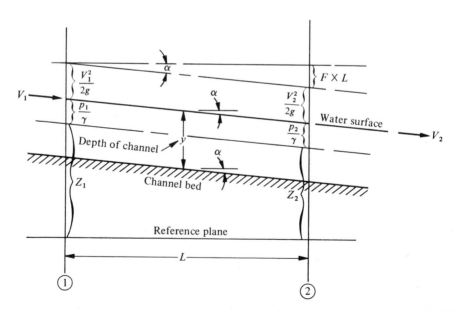

Figure 8.1. *Open channel with uniform steady flow.*

down toward the discharge end. As is shown in Figure 8.1, the channel bed has a constant slope angle of α. Since the flow is uniform and the channel has a constant depth, $V_1 = V_2$ and $V_1^2/2g = V_2^2/2g$. Therefore the water surface must be parallel to the bed of the channel and have the slope α. If the water has the specific energy $V_1^2/2g + p_1/\gamma + Z_1$ at the inlet, then it will have as its specific energy $V_2^2/2g + p_2/\gamma + Z_2 + (FL)$ at the outlet, where F is the loss in energy due to friction per unit length of channel and L the length of the channel. The slope F of the energy line is numerically equal to the slope of the free surface, which in turn is numerically equal to the slope of the channel bed. Thus these three lines (planes) must be parallel and are shown to be parallel in Figure 8.1.

With the foregoing in mind, let us consider the case of a rectangular channel with uniform steady flow. A frictional force will exist on the fluid due to the shear in the fluid layers adjacent to the channel wall. Denoting the shear stress by τ and the wetted perimeter by P, the shearing force equals τPL. If the flow is uniform, then this shearing force must equal the component of the weight of the fluid in a direction parallel to the flow. This becomes (see Figure 8.2b) equal to $(\gamma LA) \sin \alpha$, where (γLA) is simply the weight of the fluid. Equating these forces yields,

$$\tau PL = \gamma LA \sin \alpha \tag{8.1}$$

and the shear stress,

$$\tau = \frac{A}{P} (\gamma \sin \alpha) \tag{8.2}$$

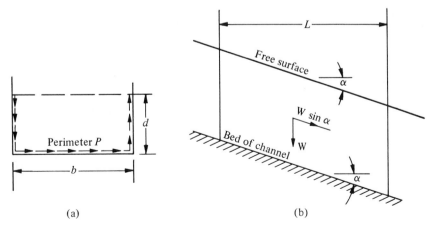

Figure 8.2. Rectangular channel.

The ratio A/P is called the hydraulic radius and is denoted by the symbol R. Therefore,

$$\tau = R\gamma \sin \alpha \qquad (8.3)$$

For small angles the sin α, the tan α, and α are all approximately equal. Thus the slope of the channel S is sin α, and we can rewrite equation (8.3) as

$$\tau = R\gamma S \qquad (8.4)$$

From our previous study of the flow in pipes, it can be stated that the shear stress will also be some function of $\rho V^2/2$. Using the mathematical symbol Ψ to denote "a function of,"

$$\tau = \Psi \frac{\rho V^2}{2} \qquad (8.5)$$

Combining (8.5) with (8.4),

$$R\gamma S = \Psi \left(\frac{\rho V^2}{2} \right) \qquad (8.6)$$

and using C to denote the resulting combined function,

$$V = C\sqrt{RS} \qquad (8.7)$$

Equation (8.7) is also known as the Chezy equation, and the student will note that the coefficient C is not a constant, and we would expect it to be a function of the Reynolds number, the relative roughness of the channel, and a factor depending on the form of the channel. Manning has empirically determined C as

$$C = \frac{1.486}{n} R^{1/6} \qquad (8.8)$$

where n incorporates the roughness of the channel. Combining (8.7) and (8.8),

$$V = \frac{1.486}{n} R^{2/3} S^{1/2} \qquad (8.9)$$

Typical values of n for various surfaces are given in Table 8.1.

Table 8.1 Flow of Water in Open Channels, Average Values of Roughness Coefficient (n), for Use in Manning's Formula

Type of Open Channel	(n)
Smooth concrete	0.014
Planed timber, asbestos pipe	0.012
Lined cast iron, wrought iron, welded steel	0.015
Vitrified sewer pipe, ordinary concrete, good brickwork, unplaned timber	0.016
Clay sewer pipe, cast iron pipe, cement lining	0.015
Riveted steel, average brickwork	0.018
Rubble masonary, smooth earth	0.025
Firm gravel, corrugated pipe	0.025
Natural earth channels (good condition)	0.025
Natural earth channels (stones and weeds)	0.032
Channels cut in rock, winding river with pools, shoals	0.040
Sluggish river (rather weedy)	0.055

The student should note the Manning's formula is empirical and is useful only over certain ranges of operation of channels. Other correlations have been proposed but these are no more accurate than equation (8.9) and in practice Manning's formula [equation (8.9)] is usually used. It should be further noted that n has the dimensions of feet$^{1/6}$ and that the derivation leading to equation (8.9) does not depend on the assumption of a rectangular channel.

ILLUSTRATIVE PROBLEM 8.1. A concrete trench, square in cross section, is to be built to carry away 2250 gal/min of water. The trench will be 1000 ft long, and the change in elevation between the ends is 10 ft. What must be the dimensions if the trench is designed to operate half full?

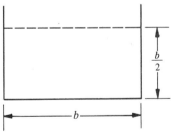

Solution. Denoting the width of the channel as b, the height of the liquid is $b/2$. The area of the channel is $b(b/2) = b^2/2$. The wetted perimeter is $(b/2) + (b/2) + b$, which is $2b$. Therefore,

$$R = \frac{A}{P} = \frac{\dfrac{b^2}{2}}{2b} = \frac{b}{4}$$

The slope S is given as 10 ft in 1000 ft or $10/1000 = 0.01$. The quantity of water flowing is given as 2250 gal/min. Since 1 gal is 231 in³., 2250 gal/min is also,

$$Q = \frac{gal}{min} \times \frac{1}{\frac{sec}{min}} \times \frac{\frac{gal}{in.^3}}{\frac{in.^3}{ft^3}} = ft^3/sec$$

$$= \frac{2250}{60} \times \frac{231}{1728} = 5 \ ft^3/sec \ flowing$$

Since

$$Q = AV, \qquad V = \frac{Q}{A} = \frac{5}{\frac{b^2}{2}}$$

where V is in feet per second, and A is in square feet. The velocity is also given by equation (8.9). Thus,

$$V = \frac{1.486}{n} R^{2/3} S^{1/2} = \frac{10}{b^2}$$

Substituting the data for this problem and using n of 0.014 as being the average value from Table 8.1, for smooth and ordinary concrete,

$$\frac{1.486}{0.014}\left(\frac{b}{4}\right)^{2/3}(0.01)^{1/2} = \frac{10}{b^2}$$

Solving,

$$b = 1.38 \ ft$$

and

$$b/2 = 0.69 \ ft$$

The flow area will be equal to $(1.38)(0.69) = 0.952 \ ft^2$ and the velocity is $10/(1.38)^2 = 5.26 \ ft/sec$.

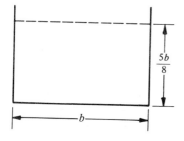

ILLUSTRATIVE PROBLEM 8.2. If the channel in illustrative problem 8.1 is designed to flow five eighths full, determine the dimensions, the velocity of the water, and the area of the channel.

Solution. Proceeding as in illustrative problem 8.1,

$$R = \frac{A}{P} = \frac{\frac{5}{8}b^2}{\frac{5}{8}b + \frac{5}{8}b + b} = \frac{5}{18}b = 0.278b$$

$$Q = 5 \text{ ft}^3/\text{sec} = AV$$

$$V = \frac{5}{\frac{5}{8}b^2} = \frac{8}{b^2}$$

Therefore,

$$V = \frac{1.486}{n} R^{2/3} S^{1/2} = \frac{1.486}{0.014} \times (0.278b)^{2/3}(0.01)^{1/2} = \frac{8}{b^2}$$

Solving,

$$b = 1.238 \text{ ft}$$
$$\tfrac{5}{8}b = 0.772 \text{ ft}$$
$$\text{Area} = 0.955 \text{ ft}^2$$

and

$$\text{Velocity} = \frac{8}{(1.238)^2} = 5.24 \text{ ft/sec}$$

Thus we see that with the same volume flow and the same slope there is a different set of dimensions that yields the same flow area and the same flow velocity.

ILLUSTRATIVE PROBLEM 8.3. If the channel designed in illustrative problem 8.1 is flowing with a depth equal to five eighths of the width, determine the flow and compare with the results of illustrative problem 8.1.

Solution. The dimensions of the channel are shown in the accompanying figure. The flow area is $(1.38)(0.863) = 1.19 \text{ ft}^2$. Proceeding as before,

$$R = \frac{A}{P} = \frac{1.19}{0.863 + 0.863 + 1.38} = 0.383 \text{ ft}$$

$$V = \frac{1.486}{n} R^{2/3} S^{1/2}$$

All items are the same as in illustrative problem 8.1 except that the hydraulic radius has changed due to the change in dimensions. Therefore we may form the ratio between the results of problem 8.1 and this problem as follows:

$$\frac{V_1}{V_3} = \left(\frac{R_1}{R_3}\right)^{2/3} = \left[\frac{\frac{1.38}{4}}{0.383}\right]^{2/3} = 0.932$$

where the subscripts 1 and 3 denote, respectively, illustrative problems 8.1 and 8.3.

$$\therefore V_3 = 1.07 V_1 = 1.07 \times 5.25 = 5.62 \text{ ft/sec}$$
$$Q = AV = 1.92 \times 5.62 = 6.7 \text{ ft}^3/\text{sec}$$

Thus increasing the area by 25 percent leads to a flow increase of 34 percent when illustrative problem 8.3 is compared to illustrative problem 8.1.

8.3 Optimum Channel of Rectangular Cross Section

In many cases it is desired to design a channel to have the maximum quantity of fluid flow for a given type of construction and with a specified slope of channel. For a fixed geometry and flow area, the maximum quantity of flow will occur when the velocity is maximum, and under these conditions equation (8.9) indicates that this occurs when the hydraulic radius is a maximum. Let us now consider a rectangular channel having a width b and a depth of water d, as is shown in Figure 8.2a. Since $R = A/P$ and A is given as a fixed quantity, R is maximum when the wetted perimeter P is a minimum. Thus we have

$$b + 2d = P \qquad (8.10a)$$

$$bd = A \qquad (8.10b)$$

Therefore,

$$\frac{A}{d} + 2d = P \qquad (8.11a)$$

If for convenience we now let $bd = A = 1$, equation (8.11a) becomes,

$$\frac{1}{d} + 2d = P \qquad (8.11b)$$

Equation (8.11b) can be solved by the methods of calculus or numerically, as is shown in Table 8.2.

By inspection of Table 8.2 (or by plotting P vs. d) it will be seen that the minimum value of P occurs when $d = \frac{1}{2}b$. In other words, the maximum velocity (or flow) for a fixed area rectangular channel of constant slope occurs when

Table 8.2

Assume d	b	d	1/d	2d	P
0.1b	3.16	0.316	316	0.632	3.792
0.2b	2.24	0.448	224	0.896	3.136
0.3b	1.83	0.549	1.83	1.098	2.928
0.4b	1.58	0.632	1.58	1.264	2.844
0.5b	1.414	0.707	1.414	1.414	2.828
0.6b	1.29	0.774	1.29	1.548	2.838
0.7b	1.195	0.837	1.195	1.674	2.869
0.8b	1.119	0.895	1.119	1.790	2.909
0.9b	1.052	0.947	1.052	1.894	2.946
1.0b	1.0	1.0	1.0	2.0	3.0

the depth of the water flowing equals half the channel width. In addition, since the perimeter is a minimum with these dimensions, the quantity of material required to construct and line the channel will also be a minimum. These conclusions do not depend on the choice of $A = 1$ in our calculations.

ILLUSTRATIVE PROBLEM 8.4. The combined discharge from two 18-in.-diameter concrete storm sewer pipes, each flowing one half full on a 1.25 percent grade, is to be carried by a cement lined open rectangular channel on a 1.0 percent grade. Design the rectangular channel so that it will have a depth of flow equal to one half of its width when carrying this discharge.

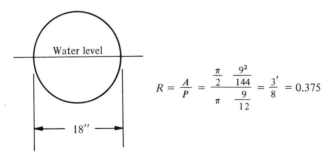

$$R = \frac{A}{P} = \frac{\dfrac{\pi}{2} \dfrac{9^2}{144}}{\pi \dfrac{9}{12}} = \frac{3'}{8} = 0.375$$

Solution. Consider the 18-in.-diameter storm sewer pipes:

$$V = \frac{1.486}{n} R^{2/3} S^{1/2} \qquad \text{from Table 8.1, } n = 0.015$$

$$= \frac{1.486}{0.015} (0.375)^{2/3} (0.0125)^{1/2} = 5.71 \text{ ft/sec}$$

Since there are two pipes,

$$\text{Total flow} = 2AV = 2\left(\frac{\pi \times (1.5)^2}{8}\right) \times 5.71 = 10.1 \text{ ft}^3/\text{sec}$$

The rectangular channel has an area of $\frac{1}{2}b^2$; therefore,

$$\frac{b^2 V}{2} = 10.1$$

and

$$b^2 V = 20.2$$

or

$$V = \frac{20.2}{b^2}$$

But

$$V = \frac{1.486}{n} R^{2/3} S^{1/2} \qquad n = 0.015 \text{ (Table 8.1)}$$

and

$$R = \frac{\dfrac{b^2}{2}}{2b} = \frac{b}{4}$$

Therefore,

$$\frac{20.2}{b^2} = \frac{1.486}{0.015}\left(\frac{b}{4}\right)^{2/3} (0.01)^{1/2}$$

$$b = 1.85 \text{ ft} \qquad \text{and} \qquad d = 0.925 \text{ ft}$$

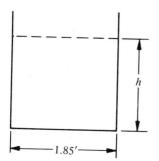

ILLUSTRATIVE PROBLEM 8.5. How much additional depth of channel in illustrative problem 8.4 will be required for the channel to carry the combined discharge when the two pipes are flowing just full?

Solution. A circular pipe flowing just full or half full has the same hydraulic radius R and therefore the same velocity V.
The total Q is therefore

$$5.71 \times 2 \times \frac{\pi \times (1.5)^2}{4} = 20.2 \text{ ft}^3/\text{sec}$$

Denoting the cement-lined channel's depth by h,

$$Q = AV = A\left(\frac{1.486}{0.015} R^{2/3} S^{1/2}\right) = 20.2 \text{ ft/sec}$$

Since $R = A/P$,

$$A \times \frac{1.486}{0.015}\left(\frac{A}{P}\right)^{2/3} (0.01)^{1/2} = 20.2$$

$$\left(\frac{A}{P}\right)^{2/3} A = \frac{A^{5/3}}{P^{2/3}} = \frac{20.2 \times 0.015}{1.486 \times 0.1} = 2.04$$

and

$$\frac{A}{P} = \frac{2.92}{A^{3/2}}$$

From the figure we also have

$$\frac{A}{P} = \frac{1.85h}{2h + 1.85}$$

Therefore,

$$\frac{1.85h}{2h + 1.85} = \frac{2.92}{(1.633)^{3/2}} = \frac{2.92}{(1.85h)^{3/2}}$$

To solve for h, let us assume various values of h and evaluate each side of the equation. Try

$h = 1$ 0.48   1.16

$h = 1.5$ 0.57   0.63

$h = 2$ 0.63   0.41

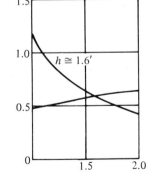

Plotting these data yields $h = 1.6$. As a check,

$$R = \frac{1.6 \times 1.85}{3.2 + 1.85} = 0.59$$

$$V = \frac{1.486}{0.015}(0.59)^{2/3} \times (0.01)^{1/2}$$

$$= 7.0 \text{ ft/sec}$$

$$Q = AV = 1.6 \times 1.85 \times 7.0 = 20.7 \text{ ft}^3/\text{sec vs. } 20.2 \text{ ft}^3/\text{sec}$$

The difference in these solutions is only 2.5 percent, which is satisfactory.

8.4 Pipes Flowing Partially Full

Quite often a large circular pipe will be used as a storm drain or sewer. In this case the pipe may not be flowing full and can be treated as an open channel with uniform flow, as was done in illustrative problem 8.4. The Chezy formula was applied to this specific situation to obtain the desired solution, but a more general treatment of this problem will be given in the following paragraphs for pipes of circular cross section.

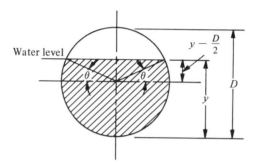

Figure 8.3

Consider the circular pipe shown in Figure 8.3, with the liquid flowing above the horizontal center line. The diameter of the pipe will be denoted as D and the depth of fluid will be denoted as y. To generalize the solution of this problem, it will be convenient to determine the ratio of the quantity of fluid flowing to the quantity when the pipe flows full. From equation (8.4), for a given pipe with a given slope,

$$\frac{Q}{Q_f} = \frac{AV}{A_f V_f} = \frac{A\left(\dfrac{1.486}{n} R^{2/3} S^{1/2}\right)}{A_f\left(\dfrac{1.486}{n} R_f^{2/3} S^{1/2}\right)} = \frac{A}{A_f}\left(\frac{R}{R_f}\right)^{2/3} \qquad (8.12a)$$

where the subscript f denotes conditions where the pipe is flowing full. Since $R = A/P$,

$$\frac{Q}{Q_f} = \left[\left(\frac{A}{A_f}\right)^{5/3}\right]\left(\frac{P_f}{P}\right)^{2/3} \qquad (8.12b)$$

From Figure 8.3, the flow area consists of the area of the sector of the circle plus the area of the triangle. In terms of the angle θ (in radians),

$$A = \left(\frac{\pi D^2}{4}\right)\left[\frac{2\left(\dfrac{\pi}{2} + \theta\right)}{2\pi}\right] + \left(y - \frac{D}{2}\right)\left(\frac{D}{2}\cos\theta\right) \qquad (8.13)$$

The wetted perimeter P is given by

$$P = \frac{\pi D \left(\frac{\pi}{2} + \theta \right)}{2\pi} \tag{8.14}$$

The angle θ is given in terms of y and D as the angle whose sine is $(y - D/2)/(D/2)$ or

$$\theta = \arc\sin \left(\frac{y - \frac{D}{2}}{\frac{D}{2}} \right) \tag{8.15}$$

For the pipe flowing full,

$$A_f = \frac{\pi D^2}{4} \quad \text{and} \quad P_f = \pi D \tag{8.16}$$

Substituting equations (8.13) through (8.16) into equation (8.12b) and simplifying,

$$\frac{Q}{Q_f} = \left\{ \left(\frac{\frac{\pi}{2} + \theta}{\pi} \right) + \frac{\cos\theta \left(\frac{2y}{D} - 1 \right)}{\pi} \right\}^{5/3} \left(\frac{\pi}{\frac{\pi}{2} + \theta} \right)^{2/3} \tag{8.17}$$

Similarly,

$$\frac{V}{V_f} = \left\{ \left(\frac{\frac{\pi}{2} + \theta}{\pi} \right) + \frac{\cos\theta \left(\frac{2y}{D} - 1 \right)}{\pi} \right\}^{2/3} \left(\frac{\pi}{\frac{\pi}{2} + \theta} \right)^{2/3} = \left(\frac{\cos\theta \left(\frac{2y}{D} - 1 \right)}{\left(\frac{\pi}{2} + \theta \right)} \right)^{2/3} \tag{8.18}$$

It will be noted from equations (8.15), (8.17), and (8.18) that Q/Q_f and V/V_f are functions only of the ratio y/D. A similar analysis also shows that Q/Q_f and V/V_f are functions of the ratio y/D when the liquid level is below the horizontal center line. These solutions are plotted in Figure 8.4, where it is seen that Q/Q_f is maximum at $y = 0.94D$ and V/V_f is a maximum at $0.8D$.

ILLUSTRATIVE PROBLEM 8.6. Water flows at the rate of 1 ft³/sec in a 24-in.-diameter storm drain having a slope of 0.0015 and n of 0.020. Calculate the depth of flow.

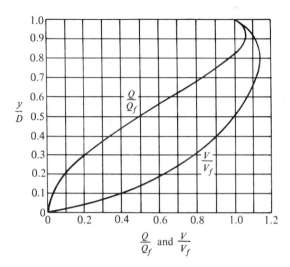

$$\frac{Q}{Q_f} \text{ and } \frac{V}{V_f}$$

Figure 8.4. *Flow in partially full circular pipe.*

Solution. If the drain is flowing full.

$$Q_f = AV = \left(\frac{\pi(2)^2}{4}\right)\frac{1.486}{0.020}\left(\frac{2}{4}\right)^{2/3}(0.0015)^{1/2} = 5.71 \text{ ft}^3/\text{sec}$$

Therefore,

$$\frac{Q}{Q_f} = \frac{1}{5.71} = 0.175$$

From Figure 8.4,

$$\frac{y}{D} = 0.28$$

Thus

$$y = 0.28 \times 2 = 0.56 \text{ ft}$$

8.5 Nonuniform Steady Flow

When a fluid flows in an open channel, the velocity is not constant over the cross section of the channel. In general there are velocity differences from side to side as well as from top to bottom of the channel. If it is desired to express the kinetic energy in terms of the mean (area-weighted) velocity, it is convenient to multiply $V^2/2g$ by a factor α, where V is the average or mean velocity. If the velocity is constant over the area, α equals unity, but depending on the character of the channel, it can equal or exceed 2. Since α cannot be predicted in advance, it is usual to assume its value to be unity; if enough information is available, it should be taken into account. In the following development α will be assumed to be unity.

Let us again consider the case shown in Figure 8.1 and now write an equation for the energy per unit weight, with the reference datum taken as the bottom of the channel. This energy quantity is known as the specific energy and is simply given by

$$E = y + \frac{V^2}{2g} \tag{8.19}$$

In the case of uniform flow, E is a constant, and for nonuniform flow, E may increase or decrease. Since $Q = AV$,

$$V = \frac{Q}{A}$$

and

$$E = y + \frac{Q^2}{2gA^2} \tag{8.20}$$

Per unit width of the channel, $A = y$, and q denotes the volume flow per unit width. Therefore,

$$E = y + \frac{q^2}{2gy^2} \tag{8.21}$$

Figure 8.5 shows a curve of E as a function of y for a fixed value of q. The 45-deg diagonal line is the potential energy for various values of y, and the horizontal distance between the 45-deg diagonal and the curve is the kinetic

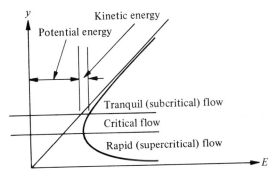

Figure 8.5. Specific energy diagram (fixed q).

energy. By inspection of Figure 8.5 it can be seen that there is a minimum value of E (at a fixed value of q) that occurs at a single value of the depth, y. This depth is the so-called critical depth, and the flow at the point is called the critical flow. For any other value of E greater than this minimum, there can physically exist two values of y. If the depth is less than the critical depth, continuity considerations require the velocity to be greater than the critical velocity, and the flow is called either supercritical or rapid flow. If y is greater than the critical depth, the velocity will be less than the critical velocity, and the flow is called subcritical or tranquil. For a rectangular channel it is not difficult to establish a relation between the minimum value of E and the value of the y for the critical depth. Using the methods of calculus it is found that

$$E_{\min} = \tfrac{3}{2} y_c \tag{8.22}$$

where $E_{\min}$ is the minimum energy and y_c the critical depth.

The foregoing considerations were for the case of constant q. If we now consider E to be constant and y and q to be variable, q is determined by solving equation (8.21). Thus

$$q^2 = (E - y)2gy^2 \tag{8.23a}$$

and for constant E, q is a function only of y. Plotting this function yields Figure 8.6. Once again it is possible to solve for the value of maximum q as a function of y, and the critical depth is found to occur at,

$$y_c = \tfrac{2}{3} E \tag{8.23b}$$

which is also the point of maximum q. Also, for a given q, the critical depth is given by

$$y_c = \left(\frac{q^2}{g}\right)^{1/3} \tag{8.24}$$

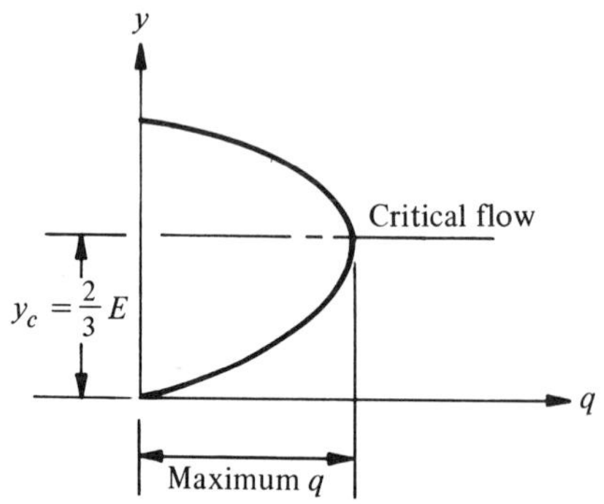

Figure 8.6. Specific energy diagram (fixed E).

In addition to these relations, it is possible to express the character of the flow in terms of the velocity and the depth. Thus from equation (8.23),

$$\frac{q^2}{y_c^3 g} = 1 \tag{8.25}$$

Since $V^2 = q^2/y^2$ or $V^2/y = q^2/y^3$,

$$\frac{q^2}{y_c^3 g} = \frac{V^2}{y_c g} = 1 \qquad (8.26)$$

Therefore, if $V^2/gy > 1$, the flow is rapid or supercritical, while $V^2/gy < 1$ is the criteria that the flow is tranquil or subcritical.

ILLUSTRATIVE PROBLEM 8.7. A channel is 10 ft wide and 5 ft deep. The rate of flow is 500 ft³/sec. Determine whether the flow is subcritical or supercritical. Also determine the second depth of flow that is possible for the same specific energy.

Solution. The quantity of flow is $AV = 500$ ft³/sec. Therefore,

$$V = \frac{500}{50} = 10 \text{ ft/sec}$$

$$\frac{V^2}{gy} = \frac{(10)^2}{32.2 \times 5} < 1.$$

The flow is thus subcritical.

From equation (8.21) (for constant E),

$$y_1 + \frac{q^2}{2gy_1^2} = y_2 + \frac{q^2}{2gy_2^2}$$

Since q is the flow per unit width,

$$q = \frac{500}{10} = 50 \text{ ft}^3/\text{sec/ft of width}$$

Therefore,

$$5 + \frac{(50)^2}{2g(5)^2} = y_2 + \frac{(50)^2}{2gy_2^2}$$

$$6.55 = y_2 + \frac{38.9}{y_2^2}$$

Simplifying,

$$y_2^3 - 6.55y_2^2 + 38.9 = 0$$

Solving numerically,

$$y_2 \simeq 3.65 \text{ ft}$$

ILLUSTRATIVE PROBLEM 8.8. A rectangular channel is to carry 300 ft³ of water/sec. Calculate the critical depth and critical velocity for a channel width of 14 ft.

Solution. Per unit width,

$$q = \frac{300}{14} = 21.4 \text{ ft}^3/\text{sec/ft of width}$$

From equation (8.24),

$$y_c = \left[\frac{q^2}{g}\right]^{1/3} = \left[\frac{(21.4)^2}{32.2}\right]^{1/3} = 2.42 \text{ ft}$$

The critical velocity

$$V_c = \frac{q}{y_c} = \frac{21.4}{2.42} = 8.85 \text{ ft/sec}$$

8.6 Hydraulic Jump

It has already been noted in Section 8.5 that at a fixed value of q there are two equally possible stable depths of flow for a given specific energy. However, it is possible under certain conditions for the character of this flow to undergo an abrupt change from one of these stable flow states to the other. Thus a channel in which the depth of flow is less than the critical depth can undergo a sudden change to a depth greater than the critical depth. In the course of this change the velocity is decreased, there is a loss in energy as the depth of flow is increased, and the flow passes through the critical state. These phenomena are known as the hydraulic jump and can be a desirable method of decreasing the velocity in a channel and converting part of the kinetic energy of the flow to potential energy beyond the jump.

To compute the height of this jump (Figure 8.7) certain simplifying assumptions will be made:

1. The frictional forces on the sides and bottom of the channel are negligibly small for the relatively short length of the jump when compared to the other energy terms.
2. The slope of the channel is assumed to be zero; i.e., the channel is horizontal.
3. The hydrostatic pressure forces are assumed to act horizontally.
4. The flow into and out of the jump is steady.
5. For this analysis we shall restrict ourselves to a rectangular channel.

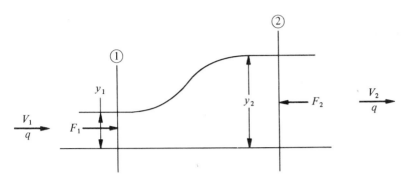

Figure 8.7. *Hydraulic jump.*

The forces F_1 and F_2 are hydrostatic forces and are simply given as the product of the respective areas multiplied by the hydrostatic pressure exerted by the fluid at the centroid of the area. Mathematically, F_1 and F_2 are given per unit width as

$$F_1 = \frac{\gamma y_1^2}{2} \quad \text{and} \quad F_2 = \frac{\gamma y_2^2}{2} \tag{8.27}$$

We can now write the equation of impulse and momentum for this problem in the direction of flow as

$$F(\Delta t) = m(\Delta V) \tag{8.28a}$$

where F is the sum of the forces in the direction of flow, m the mass undergoing acceleration per unit width of channel, and ΔV the change in velocity of the fluid in the direction of flow. Rewriting equation (8.28a),

$$F_1 - F_2 = \frac{m}{\Delta t}(V_2 - V_1) = \dot{m}(V_2 - V_1) \tag{8.28b}$$

Substituting for F_1 and F_2 and noting that $\dot{m} = \gamma q/g$,

$$\frac{\gamma}{2}(y_1^2 - y_2^2) = \frac{\gamma q}{g}(V_2 - V_1) \tag{8.29}$$

But V is simply q/y; therefore,

$$\frac{q^2}{g y_1^2} + \frac{y_1^2}{2} = \frac{q^2}{g y_2^2} + \frac{y_2^2}{2} \tag{8.30}$$

Solving equation (8.30),

$$y_1 = \frac{y_2}{2}\left(-1 + \sqrt{1 + \frac{8q^2}{gy_2^3}}\right) \tag{8.31a}$$

and

$$y_2 = \frac{y_1}{2}\left(-1 + \sqrt{1 + \frac{8q^2}{gy_1^3}}\right) \tag{8.31b}$$

From equations (8.31a) and (8.31b), y_2, the depth after the jump for a rectangular channel, is related to y_1, the depth before the jump, and the initial velocity V_1 by

$$y_2 = -\frac{y_1}{2} + \sqrt{\frac{2V_1^2 y_1}{g} + \frac{y_1^2}{4}} \tag{8.32}$$

The loss of energy $E_1 - E_2$ is given per unit mass as

$$E_1 - E_2 = \left(y_1 + \frac{V_1^2}{2g}\right) - \left(y_2 + \frac{V_2^2}{2g}\right) \tag{8.33}$$

Expressed per unit width of channel, the loss of energy in the hydraulic jump is

$$E_1 - E_2 = \left(y_1 + \frac{q^2}{2gy_1^2}\right) - \left(y_2 + \frac{q^2}{2gy_2^2}\right) \tag{8.34}$$

ILLUSTRATIVE PROBLEM 8.9. A channel is 2 ft wide and the flow is 15 in. above the base of the channel. If 15 million gal/day are flowing, what is the depth of water downstream in the channel? What is the energy loss in this process?

Solution. It is first necessary to ascertain whether the flow is greater or less than the critical flow:

$$y_c = \left(\frac{q^2}{g}\right)^{1/3}$$

Since there are 231 in.³ in 1 gal,

$$q = \frac{\dfrac{\text{gal}}{\text{day}}}{\dfrac{\text{hr}}{\text{day}} \times \dfrac{\text{sec}}{\text{hr}}} \times \frac{\dfrac{\text{in.}^3}{\text{gal}}}{\dfrac{\text{in.}^3}{\text{ft}^3}} \frac{1}{\text{channel width (ft)}} = \frac{\dfrac{\text{ft}^3}{\text{sec}}}{\text{ft width of channel}}$$

$$= \frac{15,000,000}{24 \times 3600} \times \frac{231}{1728} \times \frac{1}{2} = 11.6 \text{ ft}^3/\text{sec/ft of channel width}$$

Therefore,

$$y_c = \left[\frac{(11.6)^2}{32.2}\right]^{1/3} = 1.61 \text{ ft}$$

The depth is initially less than the critical depth and will jump to a greater depth. From equation (8.32),

$$y_2 = -\frac{y_1}{2} + \sqrt{\frac{2V_1 y_1}{g} + \frac{y_1^2}{4}}, \qquad V_1 = \frac{11.6}{\dfrac{15}{12}} = 9.27 \text{ ft/sec}$$

$$= -\frac{1.25}{2} + \sqrt{\frac{2 \times (9.27)^2 \times 1.25}{32.2} + \frac{(1.25)^2}{4}}$$

$$= -\frac{1.25}{2} + \sqrt{6.68 + 0.39}$$

$$= -0.625 + 2.65 = 2.03 \text{ ft}$$

The energy lost is

$$\left(y_1 + \frac{V_1^2}{2g}\right) - \left(y_2 + \frac{V_2^2}{2g}\right) \text{ per lb of fluid flowing}$$

But $V_2 = 11.6/2.03 = 5.71$ ft per sec:

$$\left[1.25 + \frac{(9.27)^2}{2g}\right] - \left[2.03 + \frac{(5.71)^2}{2g}\right] = 2.59 - 2.54 = 0.05 \text{ ft lb/lb}$$

The total energy lost is the total mass flowing multiplied by 0.05 ft lb/lb. Thus

$$\text{Total energy lost} = (11.6 \times 62.4 \times 0.05)(24 \times 3600) = 3,130,000 \frac{\text{ft lb}}{\text{day}}$$

$$= 0.066 \text{ hp}$$

8.7 Weirs

The quantity of fluid flowing in an open channel is usually measured by introducing a measuring device into the stream or channel. The simplest and most widely used device is a weir, which consists of an obstruction placed in the stream at right angles to the flow. As shown in Figure 8.8, the weir causes

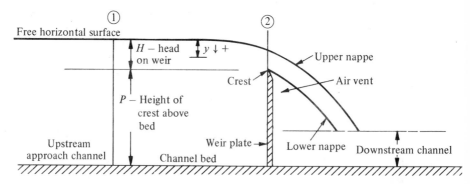

Figure 8.8. *Simplified flow over a weir.*

the stream to back up and either to flow over it or if the weir is constructed as a notch to flow through it. The notch is usually rectangular, triangular, or trapezoidal in shape and can be installed in the stream in almost any desirable manner. Some usual installations are shown in Figure 8.9. In Figure 8.9a the notch does not extend across the entire channel and is said to have two end contractions; Figure 8.9b shows a weir with one end contraction; and Figure

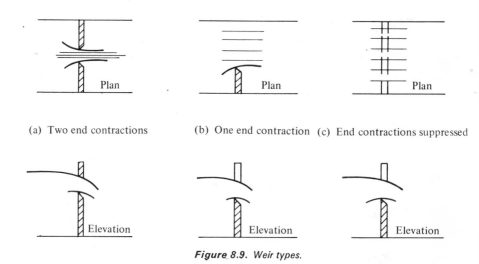

Figure 8.9. *Weir types.*

8.9c shows a weir with the notch extending the full width of the channel and having no end contractions. Since the end contractions in Figure 8.9c have been suppressed, it is commonly called a suppressed weir.

To develop a simplified analysis of a weir, the following assumptions will be made:

1. The pressure on the upper nappe and lower nappe is atmospheric. Unless this lower nappe is vented to the atmosphere, the water will tend to adhere to the lower edge of the weir.
2. The weir plate is vertical with a smooth upstream face and the flow is normal to the plate.
3. The crest is sharp and horizontal and the flow is normal to the crest.
4. Pressure losses are negligible due to the flow over the weir.
5. The channel is uniform with smooth sides upstream and downstream of the weir.
6. The approach velocity to the weir is uniform and there are no surface waves.

It is obvious that the mathematical model postulated by the foregoing assumptions does not represent the actual flow conditions in weirs. However, it does permit a rational approach to the problem of computing the flow over a weir, and the results thus obtained can be modified to conform to the experimentally determined flow.

Let us consider a rectangular suppressed weir based upon the assumptions given above, and for the present let us also impose the condition that the velocity of approach to the weir is negligibly small. Writing an energy equation between sections 1 and 2 shown in Figure 8.8 and recalling the conditions imposed by the assumptions made yields:

$$V = \sqrt{2gy} \tag{8.35}$$

where V is the velocity at a depth y in the weir. The velocity is seen to increase as $\sqrt{y}$. Let us now consider the quantity of fluid flowing over the weir. As shown in Figure 8.10, the volume of fluid flowing through the small area $(\Delta y)L$

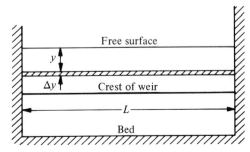

Figure 8.10. Front view of weir.

per unit of time is $(\Delta y)LV$. Since $V = \sqrt{2gy}$, the quantity of fluid flowing through this small area is $\sqrt{y}\,\Delta yL\,\sqrt{2g}$. To obtain the total flow through the weir it is necessary to sum up all of these volumes as y goes from zero to H. If $\sqrt{y}$ is plotted against y, the curve shown in Figure 8.11 is obtained. The

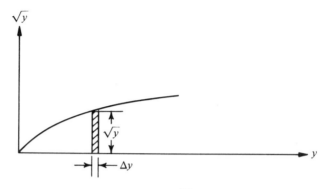

Figure 8.11. $\sqrt{y}$ vs. y.

cross-hatched area is simply $\sqrt{y}\,(\Delta y)$. Thus the total area under this curve must be proportional to the total volume flow over the weir. If this summation is performed numerically or graphically from zero to H, it will be found that the area is proportional to $H^{3/2}$. The complete solution to this problem is

$$Q = \tfrac{2}{3}\sqrt{2g}\,LH^{3/2} \tag{8.36}$$

Equation (8.36) cannot be expected to yield accurate results when applied to the flow pattern that actually exists in weirs. To account for the assumptions made in the analysis, this equation is usually multiplied by an experimentally determined coefficient C. Thus,

5.348 IMPERIAL UNITS

$$Q = C\tfrac{2}{3}\sqrt{2g}\,LH^{3/2} \tag{8.37}$$

The most common form of equation (8.37) with an empirical value for C incorporated in it is

$\therefore \rightarrow 1.83$ METRIC UNITS

$$Q = 3.33LH^{3/2} \tag{8.38}$$

which is known as the Francis formula.

ILLUSTRATIVE PROBLEM 8.10. Water is flowing through a rectangular sharp-edged weir notch 2 ft wide, 6 in. high, and set 2 ft above the bottom of an approach channel 2 ft wide and 2 ft, 6 in. deep. If the water just fills the approach channel, how much is flowing over the weir notch?

2' × 0.3048 = 0.6096 m

Solution. The accompanying figure depicts a suppressed weir, and as a first approximation the approach velocity will be neglected. Applying the Francis formula,

$$Q = 3.33LH^{3/2}$$

$$= 3.33 \times 2 \times (\tfrac{1}{2})^{3/2} = 2.36 \text{ ft}^3/\text{sec}$$

If the velocity of approach, V_1, is not negligible, the Francis formula is modified to account for this by adding a term incorporating V_1. Thus

$$Q = 3.33L\left[\left(H + \frac{V_1^2}{2g}\right)^{3/2} - \left(\frac{V_1^2}{2g}\right)^{3/2}\right] \tag{8.39a}$$

where V_1 is the approach velocity. If there are end contractions (Figure 8.9),

$$Q = 3.33\left(L - \frac{nH}{10}\right)\left[\left(H + \frac{V_1^2}{2g}\right)^{3/2} - \left(\frac{V_1^2}{2g}\right)^{3/2}\right] \tag{8.39b}$$

where n is the number of contractions of the weir. The effect of the end contractions is to make the crest shorter than the width of the channel, causing the water to contract horizontally as well as vertically in order to flow over the crest.

ILLUSTRATIVE PROBLEM 8.11. How does the approach velocity effect the flow in illustrative problem 8.10?

Solution. Using the result of illustrative problem 8.10 as a first approximation,

$$V_1 = \frac{Q}{A} = \frac{2.36}{2.5 \times 2} = 0.472 \text{ ft/sec}$$

$$Q = 3.33 \times 2\left\{\left[\left(\frac{1}{2} + \frac{(0.472)^2}{2g}\right)\right]^{3/2} - \left[\frac{(0.472)^2}{2g}\right]^{3/2}\right\}$$

$$= 3.33 \times 2\left[\left(\frac{1}{2} + 0.0023\right)^{3/2} - (0.0023)^{3/2}\right]$$

$$= 2.36 \text{ ft}^3/\text{sec}$$

It is apparent that for this case the effect of the approach velocity is negligible on the quantity of fluid flowing.

For small discharge quantities, the V-notch weir or triangular weir is widely used. Referring to Figure 8.12 and denoting the half-angle of the notch by α,

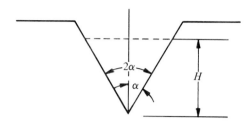

Figure 8.12. *V-Notch (triangular) weir.*

the theoretical formula for the discharge through a V notch is

$$Q = \frac{8}{15} \sqrt{2g} \, H^{5/2} \tan \alpha \qquad (8.40)$$

The actual flow quantity through V-notch weirs has been found to be closely 60 percent of the value given by equation (8.40). For 90-deg triangular weirs, the flow can be calculated from

$$Q = 2.5 H^{2.5} \qquad (8.41)$$

There are several reasons the actual discharge over a weir differs from the theoretical discharge given by equation (8.36). Among the reasons for this difference are (1) the assumption of a uniform approach velocity is not found in actual weirs; (2) the weir is not perfectly sharp and smooth; (3) viscous effects along the weir face have been neglected; and (4) the channel of approach is nonuniform. Other effects also tend to cause the actual discharge over a weir to deviate from the theoretical discharge based upon the idealized flow model that we have assumed.

Quite often it is found that the use of a weir as a measuring device is either not practical or desirable. The introduction of a weir into a channel with a very small grade may cause backing up upstream of the weir for an undesirable distance. Excessive sedimentation may occur due to the presence of the weir and the loss of head at the weir may be too great.

To overcome this problem, an instrument called the flume has been developed and is widely used throughout the world. One type of venturi flume, called the Parshall flume, has certain outstanding advantages. Among these are:

1. Installation is simple.
2. The loss in energy through the flume is very small.
3. Sedimentation in the flume is eliminated.
4. It is easily operated.

As shown in Figure 8.13, the Parshall flume provides a smooth transition

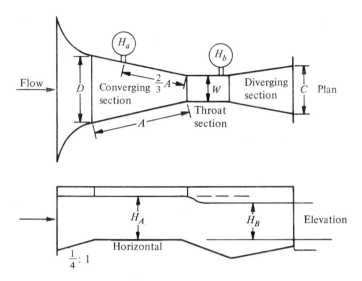

Figure 8.13. Parshall flume.

from the standard section to a reduced section and another smooth transition back to the standard section. The Parshall flume contracts the flow from the sides and bears the same relation to a sharp-edged weir that a venturi bears to an orifice in a pipe. Calibrations have been made of this flume, yielding the following equations:

$$Q = 4WH_A^{1.522W^{0.026}} \qquad \text{(for } W \text{ from 1 to 8 ft)} \qquad (8.42)$$

and

$$Q = (3.688W + 2.5)H_A^{1.6} \qquad \text{(for } W \text{ from 8 to 50 ft)} \qquad (8.43)$$

In equations (8.42) and (8.43), Q is in cubic feet per second, W is in feet, and H_A is measured as shown in Figure 8.13 in feet.

8.8 Closure

The fluid mechanics of open channel flow has been treated in a simplified manner due to the difficulty of analyzing this problem from a completely theoretical viewpoint. Our approach has been to set up a simplified model that can be treated mathematically and then to use the resulting formulation with empirically determined coefficients to make these results conform with test data. This procedure is often used in many fields and represents a workable compromise when other procedures prove inadequate. As a result of this approach we have derived formulas that apply to the flow in open channels and weirs. However, the results are limited to water as the working fluid and for a limited range of the stream conditions. In those cases where there is a need for greater accuracy then these formulas can give, it is necessary to calibrate the weir or channel.

REFERENCES

1. *Fluid Mechanics for Engineers* by P. S. Barna, Butterworth & Co. (Publishers) Ltd., London, 1957.

2. *Fluid Mechanics for Hydraulic Engineers* by H. Rouse, McGraw-Hill Book Company, Inc., New York, 1938.

3. *Basic Fluid Mechanics* by J. L. Robinson, McGraw-Hill Book Company, Inc., New York, 1963.

4. *Fluid Mechanics* by R. C. Binder, 4th ed., Prentice-Hall, Inc., Englewood Cliffs, N. J., 1962.

5. *Elementary Fluid Mechanics* by J. K. Vennard, John Wiley & Sons, Inc., New York, 1961.

6. *Mechanics of Fluids* by G. Murphy, International Textbook Company, Scranton, Pa., 1942.

7. *Fluid Mechanics for Engineers* by M. L. Albertson, J. R. Barton, and D. B. Simons, Prentice-Hall, Inc., Englewood Cliffs, N. J. 1960.

8. *Applied Fluid Mechanics* by M. P. O'Brien and G. H. Hickox, McGraw-Hill Book Company, Inc., New York, 1937.

PROBLEMS

8.1 Determine the hydraulic radius for the shapes in Figure P8.1. Assume each to be flowing full.

8.2 For each of the shapes shown in problem 8.1, water is flowing half full. Determine the hydraulic radius in each case.

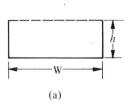
(a)

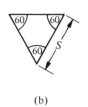
(b)

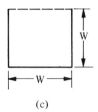
(c)

Figure P8.1

8.3 A square culvert 6 ft on each side has a slope of 1 ft in 1000 ft and is carrying water to a depth of 4 ft. If $n = 0.013$, compute the quantity of water being discharged.

8.4 A circular culvert 6 ft in diameter has a slope of 1 ft in 1000 ft and is carrying water to a depth of 4 ft. If $n = 0.013$, compute the quantity of water being discharged. Compare your result with the result in problem 8.3.

8.5 Plot $y^{1/2}$ against y for values of y from 0 to 2. Graphically determine the area under this curve and compare this to $\frac{2}{3}y^{3/2}$, where $y = 2$.

8.6 Calculate the uniform flow in a brick-lined ($n = 0.018$) trapezoidal canal whose slope is 1 ft in 10,000 ft. Assume the width at the bottom to be 5 ft and the sides to slope at 60 deg from the horizontal. The depth of flow in the canal is 6 ft.

8.7 A rectangular irrigation channel has a width of 6.0 ft, a constant slope of 0.0010, and a roughness coefficient $n = 0.011$. If the flow in the channel is uniform at 102 ft³/sec, what is the depth of water in the channel.

8.8 An irrigation canal has been provided with a bottom 20 ft wide and side slopes at 45 deg, all lined with concrete. If the canal is 10 ft deep and delivers 1050 ft³/sec of water, what is the necessary drop of level per mile? Assume that canal flows full.

8.9 A steel flume is shaped as shown in Figure P8.9. If it is flowing full and is required to carry 150 ft³/sec, what slope is required?

8.10 What are the best dimensions of a rectangular channel whose flow cross-sectional area is 150 ft²?

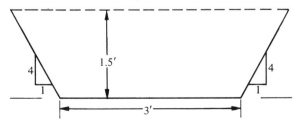

Figure P8.9

8.11 A rectangular channel is to carry 80 ft³/sec with a slope of 1.5 ft in 1000 ft. If $n = 0.015$, what are the best dimensions of the channel? Assume that the channel is flowing full.

8.12 If $n = 0.020$ in problem 8.11, determine the best dimensions of the channel. Compare the results of these problems.

8.13 A rectangular channel is made 10 ft wide. What is the critical depth of flow if 500 ft³/sec are being carried?

8.14 A rectangular channel is constructed to be 15 ft wide. At some time 2000 ft³/sec are to be carried in the channel with a velocity of 10 ft/sec. Determine whether the flow is rapid or tranquil and its specific energy.

8.15 The specific energy in a rectangular channel is 5 ft. If the channel is 4 ft wide and the flow is 25 ft³/sec, determine the possible flow depths in the channel.

8.16 If a rectangular channel is 10 ft wide, determine the maximum flow for a specific energy of 8 ft.

8.17 A rectangular channel carries a flow of 400 ft³/sec. If the channel is 8 ft wide and the water flows at a depth of 2 ft, calculate the height of the flow after a hydraulic jump.

8.18 What is the specific energy loss due to the jump in problem 8.17?

8.19 A sharp-crested weir is constructed at the end of a concrete canal 10 ft wide. The weir is 8 ft high and the channel walls extend beyond the top of the weir at a height of 9.44 ft above the bottom of the canal. The nappes are completely ventilated below the weir. How much water is flowing over the weir?

8.20 A 90 deg V-notch weir is to discharge 20 ft³/sec. Determine the head on the weir.

8.21 If the overall notch angle 2α is 60 deg in a V-notch weir, determine the head for a discharge of 20 ft³/sec. Compare the results with problem 8.20.

8.22 In connection with a water turbine test the discharge water goes into a flume 3 ft wide. At the end of this flume it is measured by a sharp created weir 3 ft high with no end contractions. If the head on the weir is 3.5 ft, what is the flow in cubic feet per second that is used in the test?

NOMENCLATURE

A = area
b = linear dimension, width
c = Chezy coefficient
D = diameter
d = linear dimension, depth
E = specific energy
F = friction loss
F = force per unit width
g = local acceleration of gravity
H = head on weir
h = height
L = length of channel
m = mass
$\dot{m}$ = mass rate of flow
n = roughness coefficient
n = number of end contractions of weir
P = wetted perimeter
p = pressure
Q = volume rate of flow
q = volume rate of flow per unit width

R = hydraulic radius = A/P
S = slope of channel
t = time
V = velocity
W = flume throat width
y = linear dimension
y = flow area per unit width
y = depth of channel
z = height
α = slope of stream bed
α = half included angle of triangular weir
γ = specific weight
Δ = small increment
θ = angle
ρ = density
τ = shear stress

Subscripts
c = critical
f = full
$1, 2, 3$ = states or locations

Chapter 9

FLOW ABOUT IMMERSED BODIES

9.1 Introduction

It is a trite truism that we on earth live at the bottom of an ocean of air. This fact must be accounted for in the design of all vehicles that move in this environment, i.e., aircraft, trains, automobiles, etc. Buildings, bridges, and other structures must also be designed to withstand the dynamic forces that this environment imposes on them. Recent advances in oceanography have opened up an entirely new field aptly called hydrospace, a relatively viscous, dense, and hostile environment in which man and his vehicles are totally immersed. The flow about an object may be due to the motion either of the object relative to the fluid or of the fluid relative to the object. In this chapter we shall consider the incompressible steady flow of a fluid relative to an object. It should also be noted that the principles developed in this chapter can also be applied to the study of turbines, fans, pumps, propellers, and many other flow systems.

9.2 General Considerations

In illustrative problem 5.4 the problem of a body moving with a velocity V while fully submerged in a viscous incompressible fluid was considered. The formulation assumed that the resisting force (called the drag force) was a function of a characteristic dimension of the body (in the case of a sphere, its diameter), the density of the fluid, the viscosity of the fluid, and the relative velocity of the fluid over the body. As a result of the analysis in Chapter 5, the following expression was developed for the drag force, where the grouping DVp/μ is the Reynolds number:

$$\frac{F}{\rho D^2 V^2} = \text{Constant} \left(\frac{DV\rho}{\mu} \right)^{-d} \tag{9.1}$$

It should be noted that the derivation leading to equation (9.1) did not include the velocity of sound (or the modulus of elasticity), and consequently we do not find the Mach number in this equation. Equation (9.1) is for incompressible flow; for compressible flow we would have to include the Mach number as a governing parameter.

If we assume that the resultant force on the body is composed of two components, one parallel to the direction of flow (called drag) and one perpendicular to the direction of flow (called lift), then equation (9.1) will basically apply to each of the components (see Figure 9.1).

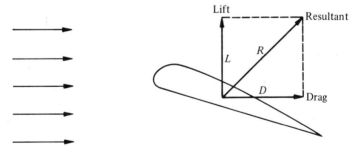

Figure 9.1. Forces on an immersed body.

It is customary to rearrange equation (9.1) by noting that D^2 is proportional to an area, A, and that $\rho V^2/2$ is proportional to the impact or dynamic pressure to obtain,

$$L = C_L \frac{\rho V^2}{2} A \tag{9.2a}$$

and

$$D = C_D \frac{\rho V^2}{2} A \qquad (9.2b)$$

where L is the lift force, D the drag force, C_L a dimensionless lift coefficient, and C_D a dimensionless drag coefficient. Both C_D and C_L can be written as functions of the single parameter, the Reynolds number as follows:

$$C_D = \varphi_1(\text{Re}) \qquad (9.3a)$$

$$C_L = \varphi_2(\text{Re}) \qquad (9.3b)$$

Equations (9.3a) and (9.3b) can be simply interpreted to mean that the lift and drag on a given body are solely functions of the Reynolds number. Figures 9.2 and 9.3 show the drag coefficient for various bodies where the area in equations (9.2a) and (9.2b) is the projected area normal to the stream.

ILLUSTRATIVE PROBLEM 9.1. In Chapter 6, Stokes' Law was discussed in relation to a sphere falling at a constant velocity in a viscous fluid. Derive a relation from Stokes' Law for C_D of a sphere for these conditions.

Solution. From Chapter 6, Stokes' Law is

$$F = 6\pi r_o \mu V$$

where ro is the outside radius of the sphere, μ the viscosity of the fluid, and V the relative velocity between the sphere and the undisturbed fluid. Equating this to equation (9.2b),

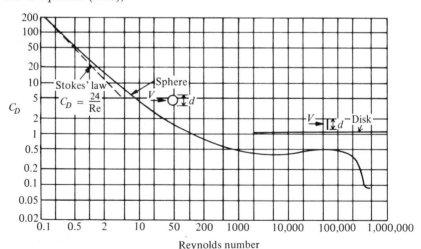

Figure 9.2. *Drag coefficients for sphere and circular disk. Reproduced with permission from* Fluid Mechanics *by R. C. Binder, 4th ed., Prentice-Hall, Inc., Englewood Cliffs, N. J., 1962, p. 168.*

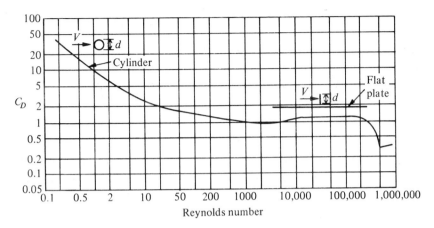

Figure 9.3. *Drag coefficients for cylinder and flat plate. Reproduced with permission from* Fluid Mechanics *by R. C. Binder, 4th ed., Prentice-Hall, Inc., Englewood Cliffs, N. J., 1962, p. 170.*

$$D = F = 6\pi r_o \mu V = C_D \frac{\rho V^2}{2} A$$

but the projected area of the sphere is πr_o^2

$$6\pi r_o \mu V = C_D \left(\frac{\rho V^2}{2}\right)(\pi r_o^2)$$

Rearranging and simplifying,

$$C_D = \frac{2(6\pi r_o \mu V)}{\pi r_o^2 \rho V^2} = \frac{24}{\dfrac{D_o V \rho}{\mu}} = \frac{24}{\text{Re}}.$$

The conclusion from this illustrative problem is that Stokes' Law yields a drag coefficient for spheres at low Reynolds numbers that is solely a function

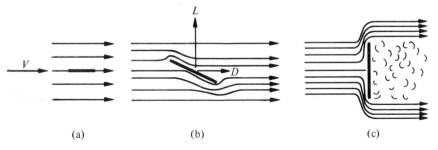

Figure 9.4. *Flat plates.*

of Reynolds number. This equation is plotted in Figure 9.2, showing the Stokes' Law line and experimental data for the sphere at various Reynolds numbers.

When a flat plate is placed parallel to the direction of fluid motion, as shown in Figure 9.4a, it experiences a drag force due to fluid friction only on both faces. As the velocity of the fluid relative to the plate is increased (with a concurrent increase in Reynolds number), the drag force decreases. At some value of Reynolds number (approximately 10^6) a sudden increase in drag occurs while a still further increase in Reynolds number yields a continual decrease in drag force. If the drag coefficient is plotted as a function of Reynolds number for smooth flat plates where the linear dimension in Reynolds number is taken to be the length of the plate parallel to the flow, the curves in Figure 9.5 are determined. These two curves appear to be similar to the curves for smooth tubes in the Moody diagram and basically are found to depend on whether the boundary layer is found to be laminar or turbulent.

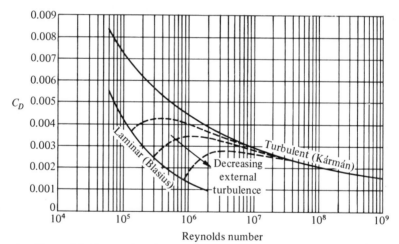

Figure 9.5. *Drag on flat plate parallel to flow direction. Reproduced with permission from* Fluid Mechanics *by R. C. Binder, 4th ed., Prentice-Hall, Inc., Englewood Cliffs, N. J., 1962, p. 173.*

When the plate is tilted at an angle to the stream, as shown in Figure 9.4b, the resultant force on the plate can be resolved into two component forces, lift (*L*) and drag (*D*). The data shown in Figure 9.6 are based upon tests on small rectangular plates whose longer side was 6 times the shorter side, with air striking the longer side first. Small corrections may be needed in applying these data to longer plates or plates of different shapes. It should be noted that these curves show no dependence on Reynolds number and that at an angle of zero degrees when the plate is parallel to the flow the drag coefficient

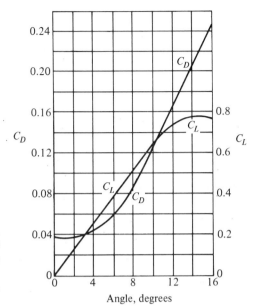

Figure 9.6. Lift and drag coefficients for a flat plate. Reproduced with permission from Elements of Practical Aerodynamics by B. Jones, 3rd ed., John Wiley & Sons, Inc., New York, 1942, p. 21.

is given as a single value. Since the conditions of the test are not specifically given and the Reynolds number is also not given, this curve should be used for design with caution. However, the qualitative trend shows an increase in both lift and drag coefficients as the angle the plate makes with the stream increases. At some finite angle, approximately 15 deg, C_L starts to decrease, but the drag coefficient C_D continues to increase.

ILLUSTRATIVE PROBLEM 9.2. Using the data shown in Figure 9.6, determine the lift and drag forces on a plate 6 ft long and 1 ft wide if the plate is set at an angle of 8 deg to an air stream having a velocity of 35 ft/sec. What is the resultant force on the plate? The specific weight of air can be taken as 0.075 lb/ft³.

Solution. Using the data from Figure 9.6 at 8 deg, $C_L = 0.51$ and $C_D = 0.086$. Therefore,

$$L = C_L \frac{\rho}{2} V^2 A = 0.51 \times \frac{0.075}{32.2 \times 2}(35)^2 \times 6 \times 1 = 4.37 \text{ lb}$$

$$D = C_D \frac{\rho}{2} V^2 A = 0.086 \times \frac{0.075}{32.2 \times 2}(35)^2 \times 6 \times 1 = 0.74 \text{ lb}$$

The resultant force equals the vector sum of L plus D. Thus

$$R = \sqrt{(L)^2 + (D)^2} = \sqrt{(0.74)^2 + (4.37)^2} = 4.44 \text{ lb}$$

Increasing the plate angle still further until the plate is perpendicular to the direction of flow yields the flow pattern shown in Figure 9.4c. For this condition the drag force is due to the pressure difference on both sides of the plate. Figure 9.7 shows the drag coefficient for finite flat plates perpendicular to the direction of flow. Comparison of the data for a square plate $(x = y)$ with the data for the disk shown in Figure 9.2 shows good agreement. As the value of x/y increases, the value of C_D approaches the value given for flat plates in Figure 9.3.

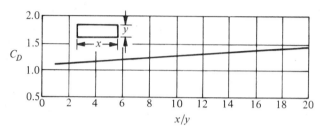

Figure 9.7. Drag coefficients; flat plate perpendicular to the flow. Reproduced with permission from Fluid Mechanics by R. C. Binder, 4th ed., Prentice-Hall, Inc., Englewood Cliffs, N. J., 1962, p. 171.

ILLUSTRATIVE PROBLEM 9.3. Compare the drag coefficient obtained when the flow of a fluid perpendicular to a flat rectangular flow is brought to rest without flow losses against one face of the plate with the data from Figure 9.7.

Solution. From the drag equation,

$$D = C_D \frac{\rho}{2} V^2 A$$

Rearranging,

$$D = \frac{C_D}{2}(\rho A V)$$

However, $\rho A V = \dot{m}$, the mass rate of flow. Therefore,

$$D = \frac{C_D}{2}\dot{m}V$$

where $\dot{m}V$ is the momentum of the fluid before it interacts with the plate. Since it is assumed that the fluid is brought to rest in an ideal manner, its momentum normal to the plate is zero after interacting with the plate. If we equate the force due to the momentum change to the drag force,

$$F = D = \dot{m}V = \frac{C_D}{2}\dot{m}V$$

and

$$C_D = 2$$

Since the fluid is not brought to rest ideally, it would be expected that C_D in the actual case would be less than 2. This can be seen to be the case by inspection of Figure 9.7.

In illustrative problem 9.1, C_D for a sphere was determined to be 24/Re based upon Stokes' Law. Stokes' Law is based upon laminar flow, and it will be seen from Figure 9.2 that for Reynolds numbers greater than 0.5, the discrepancy between values of C_D equal to 24/Re, and the experimentally determined value of C_D increases. For small Reynolds numbers, the value of C_D for various bodies is given in Table 9.1. Note that for a flat plate oriented parallel to the direction of flow the proper area to use in the drag equation is the area of both sides of the plate. It will be noted that C_D in every case is much greater than the values shown for turbulent flow on Figures 9.2 and 9.3. This was also the case for the flow of fluids in pipes, where it will be recalled the largest values of friction factor occur when the flow is laminar for small values of Reynolds number.

Table 9.1 †

Object	Re	C_D
Sphere	<0.5	24/Re
Disk flow	<0.5	20.4/Re
Disk flow	<0.1	13.6/Re
Circular cylinder	<0.1	$8\,\pi/\mathrm{Re}\,[2.0 - \ln \mathrm{Re}]$
Flat plate perpendicular to flow	<0.1	$8\,\pi/\mathrm{Re}\,[2.2 - \ln \mathrm{Re}]$
Flat plate parallel to flow	<0.01	4.12/Re

†Data for this table are based upon *Fluid Mechanics for Engineers* by M. L. Albertson, J. R. Barton, and D. B. Simons, Prentice-Hall, Inc., Englewood Cliffs, N. J., 1960, p. 395.

ILLUSTRATIVE PROBLEM 9.4. A 6-in. by 1-in. flat plate is placed parallel to the flow with the 6-in. edge facing into the flow. If the velocity of the fluid is 0.01 ft/sec, evaluate the drag on the plate. Assume that the viscosity of the fluid is 1×10^{-4} lb sec/ft^2 and that the flow is laminar.

Solution. Using $C_D = 4.12/\mathrm{Re}$ from Table 9.1,

$$\mathrm{Drag} = D = C_D \frac{\rho V^2}{2}(A)$$

Therefore,

$$D = \frac{4.12}{\frac{DV\rho}{\mu}} \frac{\rho V^2}{2} A$$

and

$$D = \frac{4.12\mu}{\frac{1}{12}V\rho} V\rho \frac{V}{2}\left[\left(\frac{1}{12}\frac{6}{12}\right)\right] \times 2$$

Note that twice the area is used for flow parallel to the plate. Rearranging,

$$D = 4.12\mu V \frac{6}{12} = 2.06\mu V$$

For this problem,

$$D = 2.06 \times 1 \times 10^{-4} \times 0.01 = 2.06 \times 10^{-6}\,\text{lb}$$

It should be noted that whenever C_D can be written as a constant/Re, the resulting equation for the drag will be a function of only the product of viscosity, the velocity, and a characteristic length, i.e., $D = $ constant $(L\mu V)$.

The previous discussion has been principally concerned with laminar flow about an immersed object. As the velocity of the fluid relative to the body is increased (with the subsequent increase in Reynolds number) we have already noted that the drag coefficient decreases. At a Reynolds number of approximately 2×10^5 it will be seen from Figures 9.2 and 9.3 that there is a sharp discontinuity, indicating a marked decrease in drag coefficient for both the cylinder and sphere. As a matter of fact the drag coefficients decrease to approximately one third of their value just prior to this occurrence. It has been found experimentally that the Reynolds number at which this abrupt decrease in drag coefficient occurs is dependent on the turbulence in the undisturbed fluid and the roughness of the body. Thus if the turbulence in the undisturbed fluid is large and/or the surface of the body is made very rough (relatively), the noted decrease in drag coefficient will occur at Reynolds numbers less than 2×10^5. This behavior is again found to be similar to the effects discussed in Chapter 6 on the incompressible flow of fluids in pipes. In both instances (flow inside pipes and flow around immersed bodies) the underlying behavior is found to reside in the boundary layer adjacent to either the pipe wall or adjacent to the immersed object. When studying the flow around an immersed body, the abrupt transition in the drag coefficient is found to be caused by the

Table 9.2 † Drag Coefficients for Cylinders and Flat Plates

Object (flow from L to R)	L/d	Re	C_D
1. Circular cylinder, axis perpendiculer to the flow	1	10^5	0.63
	5		0.74
	20		0.90
	∞		1.20
	5	$>5 \times 10^5$	0.35
	∞		0.33
2. Circular cylinder, axis parallel to the flow	0	$>10^3$	1.12
	1		0.91
	2		0.85
	4		0.87
	7		0.99
3. Elliptical cylinder (2:1)		4×10^4	0.6
		10^5	0.46
(4:1)		2.5×10^4 to 10^5	0.32
(8:1)		2.5×10^4	0.29
		2×10^5	0.20
4. Airfoil (1:3)	∞	$>4 \times 10^4$	0.07
5. Rectangular plate for which L = length and d = width	1	$>10^3$	1.16
	5		1.02
	20		1.50
	∞		1.90
6. Square cylinder		3.5×10^4	2.0
		$10^4 \times 10^5$	1.6
7. Triangular cylinder 120°		$>10^4$	2.0
			1.72
60°			2.20
			1.39
30°		$>10^5$	1.80
			1.0
8. Hemispherical shell		$>10^3$	1.33
		10^3 to 10^5	0.4
9. Circular disk, normal to the flow		$>10^3$	1.12
10. Tandem disks; spacing is L	0	$>10^3$	1.12
	1		0.93
	2		1.04
	3		1.54

†Reproduced with permission from *Fluid Mechanics for Engineers* by M. L. Albertson, J. L. Barton, and D. B. Simons, Prentice-Hall Inc., Englewood Cliffs, N. J., 1960, p. 407.

boundary layer changing from a laminar boundary layer to a turbulent boundary layer on the fore part of the body. This same phenomenon has been discussed earlier in this section during our discussion on the flow over flat plates.

Drag coefficients for cylinders and flat plates are given in Table 9.2. In the turbulent range the drag coefficient for these bodies decreases with increased Reynolds number.

9.3 Lift and Drag on Airfoils

9.3a General

In Section 9.2 the discussion was directed to the general problem of forces on an immersed body. At this point the discussion will be directed to the forces (lift and drag) that occur when an airplane moves through the air at Mach numbers $<< 1$. Prior to to our study of specific subtopics under this heading, it will be useful to define and illustrate our terminology:

Airplane. An airplane is a mechanically driven, fixed-wing aircraft, heavier than air, that is supported by the dynamic reaction of the air against its wings. This defintion can be amplified by considering an airplane in steady, level flight. For this airplane to be in equilibrium (it is not accelerating) the lift forces on it must equal its weight, and the drag forces must be countered by an equal but oppositely directed thrust from the airplane's power plant. Figure 9.8 illustrates the forces on an airplane in level flight.

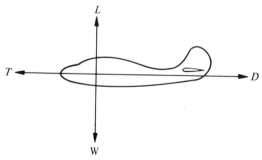

Figure 9.8. Airplane in level flight.

Airfoil. An airfoil is any surface, such as the airplane wing, aileron, or rudder, designed to obtain reaction from the air through which it moves. Figure 9.9 shows an airfoil and the pressure distribution on this airfoil. Along the upper surface there is a reduced pressure (negative pressure), while underneath the

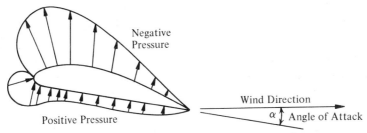

Figure 9.9. Pressure distribution on an airfoil section.

airfoil the pressure is greater than ambient (positive). This pressure distribution occurs from the acceleration of the air over the upper surface of the airfoil and the deceleration of the air over the lower surface of the airfoil. Significantly, the greater portion of the lift is obtained from the upper airfoil surface.

Angle of Attack. The angle of attack is the acute angle between a reference line in a body and the line of the relative wind direction projected on a plane containing the reference line and parallel to the plane of symmetry. The relative wind is the velocity of the air with reference to the body and it is measured at a distance from the body to minimize the disturbing effect of the body. Figure 9.9 also illustrates the angle of attack on an airfoil section.

Angle, Zero Lift. The zero lift angle is that angle of attack of an airfoil when its lift is zero. Figure 9.10 shows the characteristics of a Clark Y airfoil. With an airfoil that is completely symmetrical the lift coefficient (and consequently the lift) would be zero at zero angle of attack. However, this airfoil can be seen to be asymmetric, and it is found that C_L is zero at an angle of attack of -5 deg, i.e., that angle where the sum of the positive lift forces equals the sum of the negative lift forces. For angles of attack greater than the angle of zero lift it will be seen that the lift coefficient is directly proportional to the angle of attack when this angle is measured relative to the zero lift angle, i.e.,

$$C_L = K(\alpha - \alpha_{0L}) \tag{9.4}$$

where K is the slope of the curve of C_L against the angle of attack, α the angle of attack, and α_{0L} the angle of zero lift.

ILLUSTRATIVE PROBLEM 9.5. Using the data for the Clark Y airfoil shown in Figure 9.10, it is found that C_L is 0.5 at an angle of attack of 2 deg. Estimate C_L at an angle of attack of 6 deg.

Solution. From Figure 9.10, the angle of zero lift is -5 deg. From the stated condition at 2 deg,

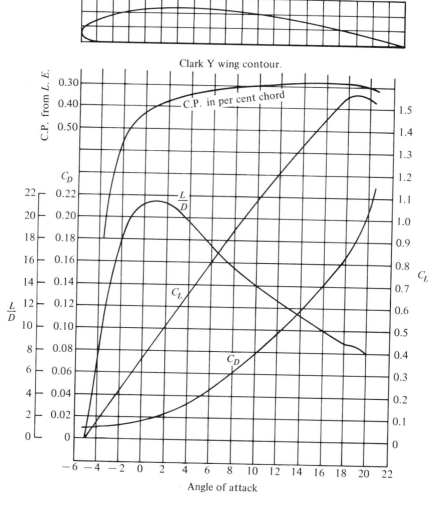

Figure 9.10. *Lift and drag coefficients for the Clark Y airfoil. Reproduced with permission from* Elements of Practical Aerodynamics *by B. Jones, John Wiley & Sons, Inc., New York, 1942, p. 39.*

$$(C_L) = K(\alpha - \alpha_{0L})$$

and

$$(0.5) = K[2 - (-5)] = 7K$$

Therefore,

$$K = \frac{0.5}{7} = 0.0714/\text{deg}$$

For 6 deg,

$$C_L = 0.0714[6 - (-5)] = 0.785$$

From Figure 9.10, C_L for 6 deg is approximately 0.79, which is in good agreement with the calculated value.

As the angle of attack is increased the lift coefficient deviates from a linear relationship with respect to angle of attack. For the Clark Y airfoil the maximum value of C_L occurs at $18\frac{1}{2}$ deg; above this angle (known as the *burble point* or *stall angle*) the lift decreases with increasing angle of attack.

Chord. The chord of an airfoil is an arbitrary datum line from which the ordinates and angles of an airfoil are measured. It is usually the straight line tangent to the lower airfoil surface at two points of the straight line joining the leading and trailing edge of the airfoil. This definition is illustrated for the two airfoils shown in Figure 9.11. The chord length (customarily given the symbol c) is the length of the projection of the airfoil profile on its chord.

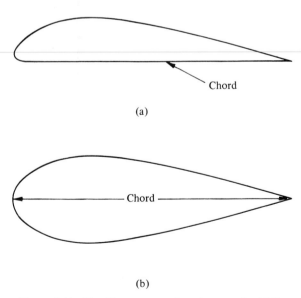

(a)

(b)

Figure 9.11. *Chord for unsymmetric and symmetric airfoils.*

Wing Area. To calculate both the lift and drag on an airplane it is necessary to know the area term in equation (9.2). For purposes of standardization the wing area is measured from the projection of the actual outline on the plane of the chords without deduction for the area that may be blanketed by the fuselage or nacelles. The wing is thus considered to extend without interruption through the fuselage and nacelles, and the wing area always includes flaps and ailerons.

Aspect Ratio (A.R.). The aspect ratio of a wing is defined as the ratio of the span of the wing to its mean chord. In general, wings are not rectangular when viewed from above, and the aspect ratio in this case is defined as the ratio of the square of the span (the distance from wing tip to wing tip is denoted as b) to the area; i.e.,

$$\text{A.R.} = \frac{b^2}{A} \tag{9.5}$$

Camber. The curvature of an airfoil is known as its camber, with upper camber referring to the upper surface, lower camber referring to the lower surface, and mean camber referring to the mean surface. It is usual to express camber as the ratio of the departure of the curved surface from a straight line joining the extremeties of the curve to the length of this straight line.

Center of Pressure (C.P). The location of the resultant aerodynamic force on an airfoil is not fixed relative to the airfoil. The center of pressure is the point in the chord of airfoil (prolonged if necessary), which is the intersection of the chord and the line of action of the resultant air force. It is practically expressed in terms of the ratio of the distance of the center of pressure from the leading edge to the chord length and is usually stated as percent of chord. Values of C.P. in percent of chord for the Clark Y airfoil at various angles of attack are shown in Figure 9.10.

Stall. When an airfoil is operated at an angle of attack greater than the angle of attack at maximum lift (for the Clark Y airfoil this is $18\frac{1}{2}$ deg), the airfoil is said to be operating in the stall condition. At stall the air separates from the airfoil or wing and forms an eddying wake with severe turbulence, and as a consequence it is found that as the stall condition is approached there is a pronounced increase in the drag coefficient. Inspection of Figure 9.10 shows the drag coefficient increasing continuously as the angle of attack is increased; the percentage increase becoming larger as the stall condition is approached.

9.3b Steady Level Flight

Let us now return to the basic lift and drag equations [equations (9.2a) and (9.2b)] and apply them to the airplane in level flight. In the following material the data given in Figure 9.10 for the Clark Y airfoil will be used as illustrative of airfoil data in general. This is not meant to infer that all airfoil data are the same or even close to that shown for the Clark Y airfoil. For other airfoils it is necessary to obtain data for the specific airfoil in question. In many instances such data will be found in the extensive compilations from wind tunnel tests conducted by NASA (and its predecessor, NACA). From equation (9.2) we can conclude the following:

1. For an airplane in steady flight the lift will just equal the weight of the airplane. Therefore,

$$W = C_L \frac{\rho}{2} V^2 A \tag{9.6}$$

Rearranging,

$$V = \sqrt{\left(\frac{W}{A}\right) \frac{2}{\rho C_L}} \tag{9.7a}$$

The term W/A is the weight loading of the wing in pounds per square foot, and therefore the velocity at which a plane must fly in steady level flight is proportional to the square root of the wing loading.

ILLUSTRATIVE PROBLEM 9.6. An airplane is to fly at 300 mph at sea level. If the airfoil is a Clark Y airfoil having an angle of 6 deg, determine the wing loading in pounds per square foot of wing.

Solution. Equation (9.7) can be rearranged using the air specific weight of 0.0765 lb/ft³ at sea level, V in miles per hour, W in pounds, A in square feet to yield

$$V = 19.77 \sqrt{\frac{1}{C_L}\left(\frac{W}{A}\right)} \tag{9.7b}$$

Using the data of this problem and $C_L = 0.79$ for an angle of attack of 6 deg,

$$300 = 19.77 \sqrt{\frac{1}{0.79}\left(\frac{W}{A}\right)}$$

and solving for W/A,

$$\frac{W}{A} = 182.5 \text{ lb/ft}^2$$

2. For an airplane in steady level flight the velocity of flight is inversely proportional to the square root of the lift coefficient. Since the lift coefficient exhibits a maximum value at the stall point, the maximum velocity (stall speed) that the airplane can have in steady level flight must occur at an angle of attack coinciding with the maximum value of C_L. For the Clark Y airfoil we have already noted this angle to be $18\frac{1}{2}$ deg. This minimum velocity will also be (very closely) the lowest landing and/or take-off speed of the airplane.

ILLUSTRATIVE PROBLEM 9.7. Using the data given in illustrative problem 9.6 and the calculated wing loading determined in this problem, evaluate the landing and take-off speed of the airplane.

Solution. At $18\frac{1}{2}$ deg, $C_L(\text{max})$ for the Clark Y airfoil is found to be closely 1.56. Therefore, using equation (9.7b)

$$V_{\text{(landing or take-off)}} = 19.77\sqrt{\frac{1}{1.56}(182.5)} = 214 \text{ mph}$$

3. When the airplane is in steady level flight the propulsion system must provide a thrust equal and opposite to the drag force on the plane. An increase in thrust while in level flight will cause the velocity of the plane to increase until the drag force once again equals the applied thrust. Conversely a decrease in thrust will cause the airplane to decrease its velocity until the drag equals the thrust. The power input by the propulsion system is the product of the thrust multiplied by the velocity of the airplane. However, in steady level flight thrust necessarily equals drag and the power required becomes

$$P = TV = DV \tag{9.8a}$$

In terms of horsepower with D in pounds and V in feet per second,

$$\text{hp} = \frac{DV}{550} = \frac{C_D \frac{\rho}{2} V^3 A}{550} \tag{9.8b}$$

where 1 hp = 550 ft lb/sec. In terms of miles per hour and a specific weight of 0.0765, we can write

$$\text{hp} = 6.8 \times 10^{-6} C_D V^3 A \tag{9.8c}$$

ILLUSTRATIVE PROBLEM 9.8. What horsepower is required for the condition given in illustrative problem 9.6? Assume that the wing has an area of 550 ft².

Solution. At an angle of attack of 6 deg, the Clark Y airfoil has $C_D = 0.045$ and applying equation (9.8c) we have,

$$hp = 6.8 \times 10^{-6} \times 0.045 \times (300)^3 \times 550 = 4540 \ hp$$

4.   The general conclusions reached in the earlier parts of this section are valid even when an airplane is operated at different altitudes. We have already concluded that steady level flight requires the lift to equal the weight of the airplane, and this must be true at all altitudes. However, if the angle of attack is constant, C_L is constant, and the velocity must increase to maintain level flight at altitudes since the density decreases with increasing altitude. Again, if the angle of attack is constant, C_D is constant, and for steady level flight it is necessary to evaluate the drag at increasing altitude since the velocity has increased but the density has decreased with increasing altitude. Let the subscript o denote sea level (zero altitude) and x any altitude; C_L and C_D are constant at all altitudes for a given angle of attack. Consider the lift equation

$$W = C_L \frac{\rho_o}{2} V_o^2 A = C_L \frac{\rho_x}{2} V_x^2 A \qquad (9.9)$$

Therefore,

$$V_x^2 = \frac{\rho_o}{\rho_x} V_o^2 \qquad (9.10)$$

From the drag equation,

$$D_o = C_D \frac{\rho_o}{2} V_o^2 A \qquad (9.11)$$

and

$$D_x = C_D \frac{\rho_x}{2} V_x^2 A \qquad (9.12)$$

Substituting (9.10) into (9.12),

$$D_x = C_D \frac{\rho_x}{2} \left(\frac{\rho_o}{\rho_x}\right) V_o^2 A = C_D \frac{\rho_o}{2} V_o^2 A \qquad (9.13)$$

The somewhat surprising conclusion from equation (9.12) is that for a given angle of attack the drag is the same regardless of altitude. However, the power required does not stay constant as can be shown by the following reasoning. From equation (9.8b) we can write

$$(hp)_o = \frac{D_o V_o}{550} \tag{9.14}$$

and

$$(hp)_x = \frac{D_x V_x}{550} = \frac{D_o V_x}{550} \tag{9.15}$$

since $D_o = D_x$.

From equation (9.10),

$$V_x = V_o \sqrt{\frac{\rho_o}{\rho_x}} \tag{9.16}$$

Substitution of (9.16) into (9.15) yields

$$(hp)_x = \frac{D_o V_o}{550} \sqrt{\frac{\rho_o}{\rho_x}} = (hp)_o \sqrt{\frac{\rho_o}{\rho_x}} \tag{9.17}$$

Since ρ_x is always less than ρ_o, the horsepower required at any altitude will be greater than the horsepower required at sea level for a fixed angle of attack.

ILLUSTRATIVE PROBLEM 9.9. A Clark Y airfoil having an area of 450 ft² is operated at a 4-deg angle of attack. If the airplane is designed to fly at 200 mph at sea level, determine its velocity at an altitude where the density is eight tenths of the density at sea level. Also determine the horsepower required at sea level and at altitude.

Solution. At 4 deg,

$$C_L = 0.65$$

$$C_D = 0.034$$

From equation (9.10),

$$V_x = V_o \sqrt{\frac{\rho_o}{\rho_x}} = 200 \sqrt{\frac{1}{0.8}} = 224 \text{ mph}$$

At sea level,

$$D_o = C_D \frac{\rho_o}{2} V_o^2 A; \qquad 200 \text{ mph} = 294 \text{ ft/sec}$$

Therefore,

$$D_o = 0.034 \frac{0.002378}{2}(294)^2 \times 450 = 1560 \text{ lb}$$

$$\frac{D_o V_o}{550} = \frac{1560 \times 294}{550} = 835 \text{ hp}$$

at altitude $(\text{hp})_x = 835\sqrt{\frac{1}{0.8}} = 935 \text{ hp}$

5.   The ability of an airfoil to perform its function of providing lift with the minimum amount of drag is expressed in the ratio L/D. Aerodynamically the wing having the largest value of this figure of merit would perform in the most desirable manner. The ratio of L/D can also be interpreted by referring to Figure 9.1, where it will be seen that the ratio of lift to drag is the tangent of the angle that the resultant force on the airfoil makes with respect to the line of the relative wind direction. Mathematically we can write

$$\frac{L}{D} = \frac{\dfrac{C_L \rho}{2V^2 A}}{\dfrac{C_D \rho}{2V^2 A}} = \frac{C_L}{C_D} \tag{9.18}$$

The ratio of L/D has been plotted in Figure 9.10 for the Clark Y airfoil. For this airfoil section it will be seen that the maximum value of L/D occurs at an angle of attack of approximately $+ 1$ deg. For a given angle of attack an airfoil having the maximum value of L/D will require less thrust from its propulsion system.

ILLUSTRATIVE PROBLEM 9.10. At $18\frac{1}{2}$ deg the Clark Y airfoil has its maximum lift and is said to be at the stall point. If an airplane using this airfoil weighs 10,000 lb, determine its drag at the stall point. Also determine the minimum drag for this airplane. Neglect all effects except those due to the airfoil. Discuss the values obtained.

Solution. At $18\frac{1}{2}$ deg,

$$\frac{L}{D} = 8.8 \quad \text{(Figure 9.10)}$$

However, $L = W$; therefore,

$$\frac{W}{D} = 8.8$$

$$D = \frac{10,000}{8.8} = 1138 \text{ lb}$$

At $+1$ deg, L/D is maximum and equal to 21.5.

$$\frac{W}{D} = 21.5$$

$$D = \frac{10,000}{21.5} = 466 \text{ lb}$$

At the higher angle of attack the required thrust is 1135 lb, which is almsot $2\frac{1}{2}$ times greater than the value at the lower angle of attack. However, it must be noted that the velocity of the airplane is not the same for both cases. Since velocity is inversely proportional to the square root of the lift coefficient, we can write

$$\frac{V_{18.5}}{V_1} = \sqrt{\frac{(C_L)_1}{(C_L)_{18.5}}} = \sqrt{\frac{0.45}{1.56}} = 0.54$$

The velocity of the airplane at $18\frac{1}{2}$ deg is therefore approximately one half the velocity that the airplane would have at 1 deg. At take-off, where it is desired that the airplane should become airborne at the minimum possible velocity (using the least length of runway), the high angle of attack is achieved using flaps and other auxiliary devices, but this is achieved only at the expense of requiring more thrust from the airplane's propulsion system.

9.3c Polar Diagram

The curves for C_L, C_D, and L/D as a function of angle of attack are satisfactory for evaluating the performance of a given airfoil, and this type of plot can also be useful when comparing airfoils. However, this requires that three curves must be plotted and evaluated in each case. It is much more convenient to plot these data on a single diagram, called a polar diagram (in which C_L is the ordinate and C_D the abscissa), as shown in Figure 9.12 for the Clark Y airfoil. If we first consider the plot to have the same scales for C_L and C_D, the ratio of lift to drag (L/D) would be the slope of the straight line drawn from the origin to the curve. The maximum value of L/D would be the slope of the line drawn from the origin tangent to the curve. In every case a line from the origin to the curve would give the direction of the resultant force and be proportional to it in magnitude. The angle of zero lift is readily obtained, and the angle of minimum drag can be found by drawing a vertical tangent to the curve. In addition, the stall point and the maximum value of C_L are also readily determined from this curve.

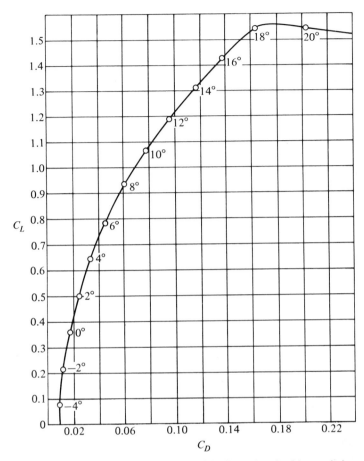

Figure 9.12. Polar curve for Clark Y airfoil. *Reproduced with permission from* Elements of Practical Aerodynamics *by B. Jones, John Wiley & Sons, Inc., New York, 1942, p. 61.*

Since values of C_D are approximately one order of magnitude less than the value of C_L at a given angle of attack, it is usual to use scales that are unequal for ease of reading, as done in Figure 9.12. When the scales are unequal, the line drawn from the origin to the curve no longer gives the direction, nor is it proportional to the magnitude of the resultant force on the airfoil. However, the tangent to the curve drawn from the origin always locates the angle of attack and the values of C_L and C_D for the maximum value of L/D.

ILLUSTRATIVE PROBLEM 9.11. From Figure 9.12 determine the stall angle, C_L, C_D, and L/D at this angle.

Solution. From Figure 9.12, the stall angle is approximately $18\frac{1}{2}$ deg,

$$C_L = 1.56, \; C_D = 0.175, \text{ and } L/D \cong 8.9$$

ILLUSTRATIVE PROBLEM 9.12. An airplane operating at sea level utilizes a Clark Y airfoil. If the airplane weighs 10,000 lb and is operated at 120 mph, determine the power required to keep it in level flight. The wing area is to be taken as 500 ft².

Solution. It is first necessary to determine the angle of attack at which the plane is operating. For this C_L is evaluated, noting 120 mph = 176 ft/sec:

$$L = W = C_L \frac{\rho}{2} V^2 A$$

$$10,000 = C_L \left(\frac{0.002378}{2}\right)(176)^2 \, 500$$

and

$$C_L = 0.544$$

From Figure 9.12, the angle of attack is $2\frac{1}{2}$ deg, and $C_D = 0.027$. Therefore, from equation (9.8c),

$$\text{hp} = 6.8 \times 10^{-6} \times 0.027 \times (120)^3 \times 500 = 159 \text{ hp}$$

9.3d Drag

If the span of an airfoil were infinite and the airfoil had an infinite aspect ratio (A. R.), it would be found to have a uniform lift across the span. For airfoils of finite length the distribution of the lift across the span is not uniform, decreasing in magnitude near the wing tips. As shown in Figure 9.9, the pressure below the airfoil is positive, while the pressure above the airfoil is negative. This pressure difference causes air on the bottom of the wing near the wing tips to flow outward and upward toward the top surface of the wing; on top the incoming air causes the airflow to move inward toward the center of the airfoil. The result of this flow pattern is to produce vortices at the wing tips, which can often be seen as an airplane passes through moist air due to condensation caused by the reduced pressure and temperature within these vortices.

As a consequence of the foregoing there is a downward flow of air at the wing tips causing changes in both the lift and drag on the wing. The magnitude of this effect is dependent on the span and aspect ratio of the airfoil; the greater the aspect ratio, the smaller is the effect on lift and drag.

Figure 9.13 shows an airfoil of finite span and the velocity of the relative wind in the undisturbed stream. For an infinite span and infinite aspect ratio

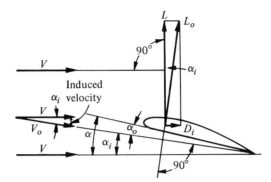

Figure 9.13. *Finite airfoil showing induced drag. Reproduced with permission from* Elementary Fluid Mechanics *by J. K. Vennard, 4th ed., John Wiley & Sons, Inc., New York, 1961, p. 522.*

there is no induced downward motion of the air so that the relative wind in the undisturbed air forward of the wing is the same as the relative wind at the wing. Due to the downward induced velocity of the finite wing, the relative wind at the airfoil does not equal the relative wind forward of the airfoil in the undisturbed air stream. If we denote the angle between the relative wind V and the true relative wind V_o to be α_i, we obtain the geometric angle of attack α from Figure 9.13 to be

$$\alpha = \alpha_o + \alpha_i \qquad (9.19)$$

where α_o is the angle of attack for an airfoil of infinite span and infinite aspect ratio. Let us now consider that the airfoil has an infinite span and infinite aspect ratio. If this were the case, α_i, the direction of the relative wind, would coincide with the line of V_o, the effective velocity, and L_o would be the lift normal to the direction of V_o. Due to the fact that L_o is not vertical, we can consider it to be composed of two components; one of these, L, is the lift directed normal to V, and the other, D_i, is called the induced drag and is directed parallel to V. Note that D_i, the induced drag, is in addition to the drag that we obtain for an airfoil of infinite span and infinite aspect ratio. Denoting the drag of an airfoil of infinite span to be the profile drag D_o, the total drag on a finite wing D can be written as the sum of the induced and profile drags:

$$D = D_i + D_o \qquad (9.20)$$

Dividing all terms in equation (9.20) by $(\rho/2)V^2A$, we obtain

$$C_D = C_{D_i} + C_{D_o} \qquad (9.21)$$

Using a mathematical approximation for the lift distribution over an airfoil of finite span, it can be shown that,

$$C_{D_i} = \frac{C_L^2}{\pi(\text{A.R.})} \tag{9.22}$$

and

$$\alpha_i = \frac{C_L(180)}{\pi^2(\text{A.R.})} = 18.24 \left(\frac{C_L}{\text{A.R.}}\right) \text{deg} \tag{9.23}$$

The importance and usefulness of the foregoing lies in the fact that airfoils of finite span that are tested in wind tunnels do not necessarily have the aspect ratio of the airfoil under consideration by the designer. In the past the aspect ratio most commonly used was 6, and by custom it can be assumed that published airfoil data are for an aspect ratio of 6 unless otherwise stated. It is presently the custom to furnish airfoil data for an infinite aspect ratio and to correct these data to any desired aspect ratio. The correction procedure consists of converting the data from one aspect ratio to an infinite aspect ratio and then reconverting to the desired aspect ratio. By this process it is necessary to report test data only for one aspect ratio (usually infinite), thereby eliminating the necessity for conducting wind tunnel tests at all aspect ratios of interest.

ILLUSTRATIVE PROBLEM 9.13. A Clark Y wing is tested at an A.R. of 6. If the lift coefficient is 0.6, determine the induced angle of attack and the induced drag coefficient and compare the induced drag coefficient to the profile drag coefficients.

Solution. From Figure 9.10, at A.R. $= 6$ and $C_L = 0.6$, the angle of attack is ~ 3.2 deg, and $C_D \simeq 0.03$. From equations (9.22) and (9.23),

$$C_{D_i} = \frac{(0.6)^2}{\pi(6)} = 0.0191$$

and

$$\alpha_i = 18.24 \left(\frac{0.6}{6}\right) = 1.8 \text{ deg}$$

$$C_{D_o} = C_D - C_{D_i} = 0.03 - 0.0191 \simeq 0.011$$

Therefore,

$$\frac{C_{D_i}}{C_{D_o}} = \frac{0.019}{0.011} = 1.72$$

The induced drag is approximately twice the profile drag.

ILLUSTRATIVE PROBLEM 9.14. A Clark Y airfoil (A.R.=6) at a 6-deg angle of attack has a C_L of 0.79 and a C_D of 0.045. Determine the lift and drag coefficients and angle of attack for a Clark Y airfoil having an aspect ratio of 8.

Solution. Since α_i is small, it is very nearly correct that $L = L_o$ and $V = V_o$ in Figure 9.13. Therefore C_L is essentially unchanged and equal to 0.79.

For the conversion of the drag coefficient it is necessary to proceed in two steps; i.e., (1) convert from an A.R. of 6 to an A.R. of ∞, and (2) convert from an A.R. of ∞ to an A.R. of 8. For the A.R. of 6 to ∞,

$$C_{D_i} = \frac{C_L^2}{\pi(\text{A.R.})} = \frac{(0.79)^2}{\pi(6)} = 0.0332$$

and

$$C_{D_o} = C_D - C_{D_i} = 0.045 - 0.0332 = 0.012 = C_{D_\infty}$$

To go from ∞ to an A.R. of 8,

$$C_D = C_{D_o} + \frac{C_L^2}{\pi(\text{A.R.})} = 0.012 + \frac{(0.79)^2}{\pi(8)} = 0.037$$

To determine the angle of attack we proceed in an analogous manner:

$$\alpha_i = 18.24\left(\frac{C_L}{\text{A.R.}}\right) = 18.24\left(\frac{0.79}{6}\right) = 2.4 \text{ deg}$$

$$\alpha_o = \alpha - \alpha_i = 6 - 2.4 = 3.6 \text{ deg}$$

For an A.R. of 8,

$$\alpha = \alpha_o + \alpha_i = 3.6 + 18.24\left(\frac{0.79}{8}\right) = 5.4 \text{ deg}$$

Therefore, at an aspect ratio of 8 and an angle of attack of 5.4 deg the lift coefficient will be 0.79 and the drag coefficient will be 0.037.

Since it is customary to present data on airfoils having an infinite aspect ratio, we can also obtain a solution to illustrative problem 9.14 using the data given in Figure 9.14 for the Clark Y airfoil of infinite aspect ratio. Illustrative problem 9.15 indicates the method of using this curve.

ILLUSTRATIVE PROBLEM 9.15. Solve illustrative problem 9.14 using Figure 9.14. Compare the results with the solution obtained for illustrative problem 9.14.

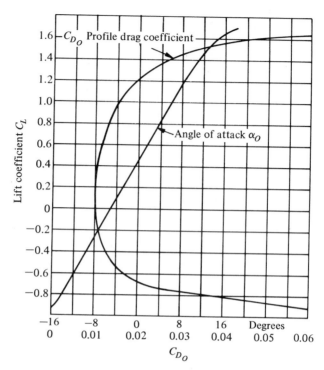

Figure 9.14. *Characteristics of a Clark Y airfoil with an infinite aspect ratio. Reproduced with permission from* Elements of Practical Aerodynamics *by B. Jones, John Wiley & Sons, Inc., New York, 1942, p. 95.*

Solution. For the infinite aspect ratio conversion to an aspect ratio of 8, it is first necessary to evaluate C_{D_o} and α_o. This is accomplished by entering Figure 9.14 at a value of C_L equal to 0.79 and reading the intersections with the curves of C_{D_o} and α_o. Proceeding in this manner we have,

$$\alpha_o = 3.6 \text{ deg}$$

and

$$C_{D_o} = 0.012$$

Since these values are essentially the same as those calculated in illustrative problem 9.14, the remainder of this problem is the same as illustrative problem 9.14.

If we combine equations (9.21) and (9.22), the resultant equation is,

$$C_D = C_{D_o} + \frac{C_L^2}{\pi(\text{A.R.})} \qquad (9.24)$$

Denoting A to be the aspect ratio of a given airfoil and B to be the aspect ratio of an airfoil of the same cross section as A but having a different aspect ratio, we can write the following

$$(C_D)_A - (C_D)_B = \left[C_{D_o} + \frac{C_L^2}{\pi(A)} \right]_A - \left[C_{D_o} + \frac{C_L^2}{\pi(B)} \right]_B \qquad (9.25)$$

However, C_{D_o} is the profile drag of an airfoil having an infinite aspect ratio and is therefore the same for both A and B. Therefore,

$$(C_D)_A - (C_D)_B = \frac{C_L^2}{\pi} \left[\frac{1}{A} - \frac{1}{B} \right] \qquad (9.26)$$

and the difference in drag coefficients for airfoils with the same airfoil cross section but differing aspect ratios can be found directly from equation (9.26).

ILLUSTRATIVE PROBLEM 9.16. Solve illustrative problem 9.14 using equation (9.26).

Solution. From equation (9.26),

$$C_{D_6} - C_{D_8} = \frac{(0.79)^2}{\pi} \left[\frac{1}{6} - \frac{1}{8} \right]$$

Therefore,

$$C_{D_6} - C_{D_8} = 0.008$$

From the given data,

$$C_{D_6} = 0.045$$

Therefore,

$$C_{D_8} = 0.045 - 0.008 = 0.037$$

which agrees with the previous calculation.

In the previous discussion the assumption has been made that C_L does not change appreciably due to the induced effects on a wing of finite aspect ratio when compared to the same wing profile having an infinite aspect ratio. Actually, as can be seen from Figure 9.15, this is not quite correct but since the value of α_i is small compared to α, this assumption is satisfactory for a first approximation.

It has already been noticed (illustrative problem 9.13) that the induced drag is a large part of the total drag on a wing of finite aspect ratio and that

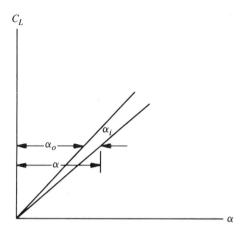

Figure 9.15. *Effect of aspect ratio on C_L.*

the induced drag of a wing with a given C_L is reduced directly as the aspect ratio is increased. For this reason, modern airplanes are built using wings that are as narrow as can be built (maximum aspect ratio) within the constraints of the requirements of the structural design and available structural materials.

9.3e Aircraft Maneuvers

The previous discussion has considered only the airplane in steady level flight. A complete discussion of the mechanics of unsteady flight is beyond the scope of this text, but we have already developed enough information to consider certain steady-state aircraft maneuvers, namely, (1) glide, (2) banked turn, and (3) climb.

When an airplane is said to be in a glide we shall assume that the power is off (engine completely throttled or dead) and that the propulsion system does not provide any thrust or drag. The glide angle is defined as the angle below the horizontal that the airplane descends along under these conditions. Denoting this angle as θ, the angle of incidence to be α, and the lift, drag, and weight as L, D, and W, respectively, we obtain the free-body diagram shown in Figure 9.16. In this figure the lift is perpendicular to the flight path, and the drag is parallel to the flight path. For a steady glide, the lift, drag, and weight must be a system of concurrent forces in equilibrium, and from Figure 9.16 and the conditions of equilibrium we can write

$$L = W \cos \theta \tag{9.27}$$

but

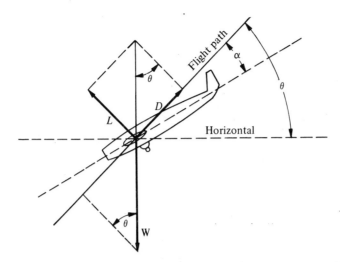

Figure 9.16. *Airplane in a glide.*

$$L = C_L \frac{\rho}{2} V^2 A \tag{9.28}$$

Therefore,

$$\cos \theta = \frac{L}{W} = \frac{C_L \frac{\rho}{2} V^2 A}{W} \tag{9.29}$$

However,

$$D = W \sin \theta \tag{9.30}$$

and

$$W = \frac{D}{\sin \theta} = \frac{C_D \frac{\rho}{2} V^2 A}{\sin \theta} \tag{9.31}$$

Combining equations (9.29) and (9.31),

$$\tan \theta = \frac{D}{L} = \frac{C_D}{C_L} \tag{9.32a}$$

or

$$\cot \theta = \frac{L}{D} = \frac{C_L}{C_D} \tag{9.32b}$$

The lift to drag ratio, L/D, can be read directly from curves such as Figures 9.10 or 9.12 for a given airfoil. Note that for each angle of incidence there is a single value of L/D, and therefore there is a single value of the glide angle. Also, since the ratio of lift to drag (and consequently θ) is not a function of the air density, the glide angle is the same for all altitudes.

ILLUSTRATIVE PROBLEM 9.17. An airplane using a Clark Y airfoil of A. R. 6 has a power plant failure and must glide to a nearby airport. Assuming the airplane to be at an altitude of 1000 ft, can the plane glide to the airport if the airport is 2 miles away?

Solution. The maximum horizontal distance that the plane can glide occurs when the plane is operated at the minimum glide angle, and this corresponds to a maximum value of L/D. For the airplane in question, Figure 9.10 indicates a maximum value of L/D to be closely 21.5 at an angle of attack of approximately 1 deg. The $\tan \theta = 1/21.5 = 0.0465$ and $\theta \simeq 2.7$ deg. Referring to the accompanying figure, the maximum distance the plane can glide is found as follows:

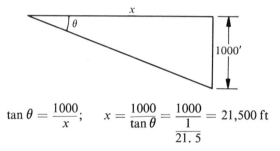

$$\tan \theta = \frac{1000}{x}; \quad x = \frac{1000}{\tan \theta} = \frac{1000}{\dfrac{1}{21.5}} = 21{,}500 \text{ ft}$$

or 4.08 miles. Therefore it is possible for the plane to safely glide to the airport.

As the angle of the glide is made steeper and steeper, a vertical "dive" occurs when the glide angle is made 90 deg. Under this condition the lift is zero, the angle of attack corresponds to the angle of zero lift, and the plane will reach a velocity, termed the terminal velocity, V_T, of

$$V_T = \sqrt{\frac{2W}{C_D \rho A}} \tag{9.33}$$

The terminal velocity is seen to decrease as the altitude decreases since the air density increases with decreasing altitude. If the airplane is placed in a dive (either deliberately or otherwise) with power on, the thrust of the airplane's power plant must be accounted for. For this condition the drag on the airplane must equal the sum of the weight and thrust; thus

$$T + W = D = C_D \frac{\rho}{2} V^2 A \qquad (9.34)$$

and the terminal velocity with power is given by

$$V_{TP} = \sqrt{\frac{2(W + T)}{C_D \rho A}} \qquad (9.35)$$

Depending on the altitude and thrust it is possible for the plane to continue to accelerate and never reach its terminal velocity. Unless the airplane's controls can be made to bring it out of the dive without structural damage, it will crash and be destroyed.

ILLUSTRATIVE PROBLEM 9.18. Due to a malfunction in the controls of an airplane, it is placed in a dive. If the thrust is 2000 lb, the weight is 3000 lb, and the airfoil is a Clark Y airfoil of A.R. 6, determine the terminal velocity at sea level if the wing area is 600 ft².

Soluiton. For the Clark Y airfoil of A.R. 6, the angle of zero lift is obtained from Figure 9.10 as approximately −5 deg. C_D corresponding to this angle of attack is closely 0.01, and the terminal velocity will be

$$V_{TP} = \sqrt{\frac{2(2000 + 3000)}{0.01(0.002378)600}} = 836 \text{ ft/sec}$$

or

$$V_{TP} = 570 \text{ mph}$$

When an airplane in steady level flight changes its flight direction by turning, it is necessary to consider the centrifugal force acting during the turn. As the airplane is turned in a horizontal plane it is usual to "bank," that is, to depress the inner wing and elevate the outer wing. Without banking, the airplane tends to move outward; with banking the motion of the airplane depends on the angle of bank, the velocity of the airplane, and the radius of the turn. Consider the free-body diagram of the banked airplane shown in Figure 9.17. For equilibrium, the centrifugal force WV^2/gR must be balanced by the horizontal component of the lift, and the weight must be balanced by the vertical component of the lift. Thus

$$L \sin \theta = \frac{WV^2}{gR} \qquad (9.36)$$

and

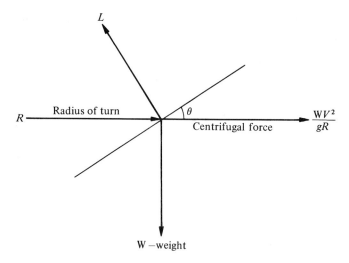

Figure 9.17. *Forces in a steady banked turn.*

$$L \cos \theta = W \tag{9.37}$$

From equations (9.36) and (9.37),

$$\tan \theta = \frac{V^2}{gR} \tag{9.38}$$

The angle of bank is thus seen to be independent of the weight of the airplane, the wing area, and the properties of the airfoil. If the angle of the bank is made less than the value given by equation (9.38), the plane will move outward and is said to skid. If the angle of bank exceeds the value given by equation (9.38), the airplane will move inward and downward and is said to slip.

ILLUSTRATIVE PROBLEM 9.19. An airplane in a race over a closed course is required to turn around a pylon at each end of the course. If the plane is traveling at 400 mph and is restricted to a turn radius that must not exceed 1 mile, determine the best bank angle.

Solution. For proper banking without skidding or slippng,

$$\tan \theta = \frac{V^2}{gR}$$

$V = 586 \text{ ft/sec}$ and $R = 5280 \text{ ft}$.

$$\tan \theta = \frac{(586)^2}{32.2(5280)} = 2.02$$

$$\theta = 63.7 \text{ deg}$$

The plane must make a very steep bank if it is to perform within the required limitations.

ILLUSTRATIVE PROBLEM 9.20. If the pilot in illustrative problem 9.19 weighs W lb, what is his apparent weight in the turn?

Solution. The total force experienced by any object in the airplane is the vector sum of its weight and the centrifugal force acting on the body. Therefore,

$$\text{Total force} = \text{Apparent weight} = \sqrt{\left(\frac{V^2}{gR}\right)^2 + W^2}$$

Factoring,

$$\text{Apparent weight} = W\sqrt{\left(\frac{V^2}{gR}\right)^2 + 1} = W\sqrt{\tan^2\theta + 1} = \frac{W}{\cos\theta}$$

The ratio of the apparent weight to W is the g loading, i.e., the force the body would experience in a gravitational field greater than earth's gravitational field by some multiple of the earth's gravitational field. Therefore the g load is

$$g \text{ load} = \frac{\text{Apparent weight}}{W} = \frac{1}{\cos\theta} = \sec\theta$$

For the problem under consideration, $\theta = 63.7$ deg and $\cos\theta = 0.443$:

$$g \text{ load} = \frac{1}{0.443} = 2.26$$

Therefore the apparent weight of the pilot is $2.26W$. Notice that the g load is a function only of the bank angle, θ.

When an airplane is in steady level flight, the lift force must equal the weight of the ariplane. If it is desired to climb from one altitude to another, the lift provided must be in excess of that required for steady level flight, and the additional power must be provided by the airplane's propulsion system. In effect, work is done lifting the airplane from one altitude to another. The work done in lifting an airplane of W lb from one altitude (H) to another (H_2) is

$$\text{Work} = W(H_2 - H_1) \tag{9.39}$$

If both sides of equation (9.39) are divided by t, the time it takes to go from H_1 to H_2,

$$\frac{\text{Work}}{t} = \text{Power to climb} = \frac{W(H_2 - H_1)}{t} \tag{9.40}$$

The term $(H_2 - H_1)/t$ is the vertical velocity in the climb or the rate of climb (R.C.); the rate of climb is then

$$\text{R. C.} = \frac{\text{Power to climb}}{W} \tag{9.41}$$

At any altitude the airplane's propulsion system can generate a maximum amount of power, which we can denote as power available, and we can denote the power required for level flight at that altitude as power required. Clearly, the extra power required for the climb must come from the difference between the power available and power required, and the maximum rate of climb will occur when the power to climb is equal to the difference between the power available and the power required. We can incorporate this into equation (9.41) and arrive at equation (9.42), where R.C. is the rate of climb in feet per second, W is the weight in pounds, and hp denotes horsepower:

$$\text{R. C.} = \left(\frac{\text{hp available} - \text{hp required}}{W}\right)550 \tag{9.42}$$

ILLUSTRATIVE PROBLEM 9.21. An airplane weighing 10,000 lb is designed to climb at the rate of 300 ft/sec. Determine the excess power that must be provided by the propulsion system.

Solution. From equation (9.42),

$$300 = \frac{(\text{Excess power})(550)}{10,000}$$

and

$$\text{Excess power} = \frac{300 \times 10,000}{550} = 5450 \text{ hp}$$

9.4 Closure

In this chapter we have discussed the flow about immersed bodies from both general and applied aspects. For a subject as vast as this particular topic, it would indeed try anyone to attempt more than a survey of selected portions of this subject within the limits of a single chapter in an introductory text. However, we have been able to discuss and differentiate between the laminar and turbulent flow regions for flow about immersed bodies. We have also been able to study the airfoil in compressible subsonic flow and relate the performance of the airplane to the properties of the airfoil. Wherever possible the development was based upon reasonable mathematical models, but it was found necessary ultimately to resort to experimentally determined characteristics for each of the topics covered.

This chapter is intended to be an introduction to selected areas of aerodynamics and hydrodynamics. Mastery of it will provide a sound basis for further study in these areas.

REFERENCES

1. *Fluid Mechanics* by R. C. Binder, 4th ed, Prentice-Hall, Inc., Englewood Cliffs, N. J., 1962.

2. *Fluid Mechanics for Engineers* by M. L. Albertson, J. R. Barton, D. B. Simons, Prentice-Hall, Inc., Englewood Cliffs, N. J., 1960.

3. *Elementary Fluid Mechanics* by J. K. Vennard, John Wiley & Sons, Inc., New York, 1961.

4. *Elements of Practical Aerodynamics* by B. Jones, John Wiley & Sons, Inc., New York, 1942.

5. *Basic Fluid Mechanics* by J. L. Robinson, McGraw-Hill Book Company, Inc., New York, 1963.

6. *Mechanics of Fluids* by G. Murphy, International Textbook Company, Scranton, Pa. 1942.

7. *Engineering Applications of Fluid Mechanics* by J. C. Hunsaker and B. G. Rightmire, McGraw-Hill Book Company, Inc., New York, 1947.

PROBLEMS

Use

$$\mu = 3.73 \times 10^{-7} \, \text{lb sec/ft}^2$$

$$\rho = 0.002378 \frac{\text{lb sec}^2}{\text{ft}^4} \qquad (\gamma = 0.0765 \, \text{lb/ft}^3)$$

for air in the following problems unless otherwise noted.

9.1   A sphere whose specific weight is 100 lb/ft³ settles in air at a velocity of 0.3 ft/sec. Evaluate the diameter of the sphere. Neglect the buoyant effect of the air.

9.2 If a sphere is immersed in a fluid whose relative velocity with respect to the sphere is 0.01 ft/sec, determine the drag force on the sphere. Assume the Reynolds number to be less than 0.5, the viscosity of the fluid to be 3×10^{-5} lb sec/ft^2, and the diameter of the sphere to be 0.001 in.

9.3 Using the data for cylinders in Figure 9.3, derive a relationship for the drag coefficient on a cylinder for Re $\leq$ 0.5 of the form $C_D = $ constant/Re.

9.4 Using the results of problem 9.3, solve problem 9.2 for a cylinder placed normal to the flow and having a length of 1 in. This is essentially a short fine wire having a large ratio of length to diameter.

9.5 What force is exerted on a billboard 10 ft long and 5 ft high by a 100-mph wind perpendicular to it?

9.6 Air flows edge-on to a rectangular plate 2 × 1 ft. Determine the drag on the plate if the air velocity is 30 mph and the stream strikes the 2-ft side first. Assume the flow to be laminar.

9.7 If the plate in problem 9.6 is to be towed through the air at 30 mph, what horsepower is required?

9.8 What is the drag on a spherical ball 3 in. in diameter on top of a flag pole when the relative wind velocity is 60 mph?

9.9 Determine the ratio of the drag on a flat plate for angles of the relative wind of 16 and 8 deg.

9.10 A plate 1 ft long and 8 in. wide is set at an angle of 12 deg to an air stream with a relative velocity with respect to the plate of 75 ft/sec. Determine the lift, drag, and total force on the plate.

9.11 A hemispherical shell is placed in a fluid so that the relative velocity of the fluid impinges on the circular plane face. Assume the hemisphere to be 4 in. in diameter, the velocity of the fluid to be 10 ft/sec relative to the hemisphere, the Reynolds number to be greater than 1000, and the fluid to have a specific weight of 100 lb/ft^3. For these conditions determine the drag on the hemisphere.

9.12 If the hemisphere in problem 9.11 is turned through an angle of 180 deg, what will the drag be?

For the following problems, unless otherwise stated assume a Clark Y airfoil of A.R. 6.

9.13 What is the lift, drag, and ratio of lift to drag at an angle of attack of 9 deg on a wing having an area of 400 ft^2 at a relative velocity of 180 mph? How much horsepower is required?

9.14 An airplane has a wing load (W/A) of $100 \, lb/ft^3$ and is operated at $240 \, mph$. What is its angle of attack?

9.15 What wing area is required to support an airplane weighing $500 \, lb$ and flying at an angle of attack of 7 deg with a velocity of $180 \, mph$?

9.16 What wing loading will an airplane have if it operates at an angle of attack of 4 deg at $200 \, mph$?

9.17 If the plane in problem 9.16 operates at an altitude where the density of the air is half that at sea level, determine the permissible wing loading.

9.18 Determine the landing speed of the airplane in problem 9.14.

9.19 An airplane takes off from New York (sea level) and is to land at Denver (altitude, 7500 ft). The specific weight of air at Denver is approximately 0.8 times the value at New York. Determine the ratio of the take-off speed from New York to the landing speed at Denver.

9.20 Using the data given in problem 9.19, determine the ratio of the horsepower for take-off from New York to the horsepower required for take-off from Denver. Assume the same angle of attack for each case.

9.21 It is desired to operate an airplane at the angle of maximum lift. If the plane weighs $6000 \, lb$ and has an area of $200 \, ft^2$, determine the operating velocity, the drag, and horsepower required. Determine the same items if the plane is operated at an angle of attack of 14 deg. Compare both modes of operation.

9.22 The largest engine an airplane can utilize generates $1000 \, hp$. At what angle of attack should the plane be operated with this engine if the design speed is $300 \, mph$ and the wing area is $200 \, ft^2$?

9.23 An airplane operates at a condition such that $C_L = 0.8$ and the drag is $500 \, lb$. What is the lift?

9.24 An airplane has a span of $75 \, ft$ and a mean chord of $5 \, ft$. If $C_L = 0.8$, what is the induced drag coefficient?

9.25 For problem 9.24 what is the induced angle of attack?

9.26 If a Clark Y airfoil (A.R. $= 6$) is operated at $C_L = 0.5$, determine the drag coefficients and the angle of attack for a Clark Y airfoil having an A.R. of 9 at the same value of C_L.

9.27 What is the ratio of the induced to total drag on a Clark Y airfoil having an A.R. of 6 and an angle of attack of 6 deg?

9.28 An airfoil has an infinite aspect ratio. At a given angle of attack $C_D = 0.03$ and $C_L = 1.0$; what is C_D for a similar airfoil of A.R. 6 when $C_L = 1.0$?

9.29 Determine α and C_D for an airfoil of A.R. 8 and the same C_L as a similar airfoil of A.R. 6 having $C_L = 0.7$.

9.30 What is the least glide angle for a wing having an area of 450 ft² and weighing 2500 lb?

9.31 Evaluate the terminal velocity of the airplane in problem 9.30.

9.32 If the airplane in problem 9.30 has a thrust of 2000 lb, determine its terminal velocity in a power dive.

9.33 If the airplane in problem 9.30 is at 750 ft when the glide is started, determine the farthest distance the plane can glide.

9.34 What angle of bank is required for an airplane traveling at 240 mph in a horizontal turn of radius equal to 1 mile?

9.35 Evaluate the g force and radius of turn for an airplane weighing 3500 lb that is turning at a bank angle of 45 deg with a velocity of 180 mph.

9.36 A pilot of a racing airplane places it in a 75-deg banked turn. Determine the g force on the pilot.

9.37 At a given altitude, an engine develops 1000 hp while the airplane requires 500 hp to fly in steady level flight. Determine the maximum rate of climb this plane can have under these conditions if it weighs 10,000 lb.

NOMENCLATURE

A = area
A = aspect ratio A
A.R. = aspect ratio
B = aspect ratio B
b = span
C_D = drag coefficient
C_L = lift coefficient
C.P. = center of pressure
c = chord length
D = linear dimension
D = drag
D_i = induced drag
D_o = profile drag
d = linear dimension
d = dimensionless exponent
F = force
g = acceleration of gravity
H = altitude
hp = horsepower
K = slope
L = length
L = lift
$\dot{m}$ = mass rate of flow
P = power
R = resultant force

R.C. = rate of climb
Re = Reynolds number
r_o = outer radius of sphere
T = thrust
t = time
V = velocity
W = weight
x = width of plate
y = height of plate
α = angle of attack
α_i = induced angle of attack
α_o = angle of attack, infinite A.R.
θ = angle
μ = viscosity
ρ = density

Subscripts

o = sea level
TP = terminal with power
x = any altitude
$0L$ = zero lift
$1, 2$ = different states
i = induced
o = profile

Chapter 10

FLUIDICS[1]

10.1 Introduction

In recent years the application of principles of fluid mechanics, long recognized but not used, has led to the new technology called fluidics. Fluidics can be termed to be the technology of using streams of gas or other fluids to perform such logic functions and control functions as amplification, sensing, logic switching, computation, and control. As a new technology, fluidics offers substantial technical advantages and in some instances has proved itself to be an inexpensive, reliable substitute for electronics. In other cases the combination of fluidics with either mechanical or electronic components has led to a hybrid system having characteristics that could not be obtained in any other manner. The advantages of fluidic devices can be categorized as follows:

1. *Reliability.* Simplicity of construction is a characteristic of all fluidic devices, and since these devices have no moving parts with the exception of the

---

[1]The material in this chapter has been taken principally from the following sources with permission: (1) "Fluidics: Development and Outlook" by Alexander Block, *Engineer*, **X**, No. 1, January/February 1969, p. 14; (2) "Fluidics" by Stanley M. Shinners, *Electro-Technology (New York)*, **79,** March 1967, p. 81.

fluid used, they are inherently reliable. Fluidic devices can be made of a variety of materials including glass and plastics, and by the proper selection of fluid and structural material the design can virtually eliminate chemical reactions. Although there are insufficient data available at this time to be able to statistically predict the reliability of fluidic units, it has been predicted that improvements of two orders of magnitude can be expected for fluidic components over comparable electronic components.

2. *Environmental.* As previously noted, fluidic devices can be fabricated from any material and almost any fluid can be used as the operating fluid. The only restriction on the operating fluid is the possible effect that internal contamination may have since the need for filtration to keep the working medium clean increases both the initial and operating costs of a fluidic system. When the proper material is chosen for a particular environmental situation, fluidic devices can be used where it would be impossible to use electronic units. Shock and vibration tests have been run on fluid devices from 0 to 5000 Hz at accelerations to $50\,g$ without any indication of malfunction. Also, if a suitable ceramic base material is used, operation at temperatures to $3000°$ or $4000°F$ may be possible. Additionally, fluidic devices do not generate, nor are they affected by, radiation. This attribute eliminates the costly radio-frequency (rf) shielding techniques required by electronic systems. The fact that they are not affected by radiation makes fluidic devices particularly applicable to space equipment. The environmental reliability of fluidics was studied as early as 1962, when researchers reported that failures induced by temperature change were reduced 5 times when fluidic devices were substituted for equivalent electronic units. Nuclear radiation failures were reduced by a factor of 6, as was vibration failure.

3. *Safety.* Very important, too, is the safety claimed for fluidic devices. Inasmuch as they replace electronic or electrical controls, they eliminate all shock and explosion hazards. Absence of electrical wiring precludes short circuits. The use of low-pressure noncorrosive operating fluids also increases the safety of fluidic devices.

4. *Cost.* Fluidic devices can be fabricated by any one of several low-cost manufacturing methods: stamping, coating, injection molding, or etching. Units with interconnecting holes in the proper locations can be stacked on top of each other to achieve compact, inexpensive systems. It has been estimated that the cost of producing a particular fluid amplifier using these techniques would be cheaper than an equivalent electronic unit.

The disadvantages of fluidic devices are:

1. *Speed of response.* At present, fluidic devices suffer from relatively low speeds. Also, only low pressure is used at this time, and thus only low-power output levels can be attained. Direct actuation of powered equipment is, in most cases, not possible, which requires amplification. This is expensive and may negate to some extent the reliability and simplicity of fluidic devices. The reasons for the low speeds and pressure lie in the fact that almost all these devices are pneumatic. The velocity of gases going through ordinary nozzles is practically limited to the speed of sound in the particular gas. In air, the usual medium, sonic choking of a nozzle occurs at approximately twice the vent pressure. And since almost all fluidic devices are vented to atmosphere, the upper pressure limit at present is 35 psi. It may be possible to go to supersonic flow and thus increase both velocity and pressure. However, considerable basic research is required to understand shock wave phenomena and to design suitable equipment.

Incompressible fluids, such as hydraulic oil, have been attempted in fluidic devices but with minor success. At low pressures, liquids are not economical, while at higher pressures, even only 100 psi, cavitation occurs, with resulting nozzle erosion.

2. *Cross-coupling effects.* Problems in making interconnections are often encountered, especially in applications where a great number of interlocking functions are required. The interconnection of fluid elements is very complex compared to the wiring of electronic circuits because the interconnecting tubes act as transmission lines. The lack of a unified theoretical approach increases the problem.

There are two methods of categorizing fluidic devices: (1) by the type of fluid flow interaction occurring in them or (2) by their funcion. For example, using the flow interaction method of classification it is possible to use the general categories of wall attachment, momentum exchange, turbulence, and vortex to describe various fluidic devices. By function these same devices fall under the categories of amplifiers, sensors, and logic elements. Sensors are used to detect some physical condition or change and provide a fluid signal as an output; amplifiers (or logic elements) are used to build up this signal and process or store it; actuators are then instructed to act on the processed signal and, if desired, perform some physical action. In this chapter a combination of both systems will be used to describe the technology of fluidics.

10.2 Coanda Effect

Henri Coanda, a Rumanian engineer, found that a fluid jet attaches itself
to one wall of a flat nozzle and stays there because of a stable, dynamically
formed and sustained pressure gradient across the stream. Figure 10.1 illustrates
the attachment of a jet to a wall due to the Coanda effect. The design of many
fluid amplifiers is based upon this Coanda effect and the momentum-transfer
principle, which is based upon the principle that a jet of fluid emerging from
a nozzle can be deflected by another perpendicularly directed jet of much lower energy, as will be discussed later in this chapter. Figure 10.2 illustrates the

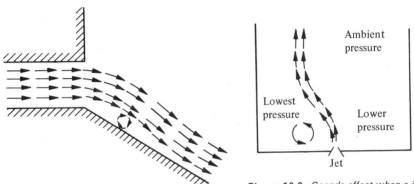

Figure 10.1. Coanda effect.

Figure 10.2. Coanda effect when a jet
enters a region with two side walls.

Coanda effect on a fluid jet entering a region containing two side walls. When
there are side walls near the jet, the entrainment of ambient air into the mix-
ing zones of the jet reduces pressure between the jet and the side walls. Because
of the random characteristics of the turbulence of the jet, it wanders back and
forth toward the side where the lower pressure exists, eventually touching the
wall. Since the region between the wall and the jet is cut off from supply by
the ambient atmosphere, its pressure becomes even lower. The result is that
the jet stream becomes stable in the attached position. This is the basic Coanda
principle as it applies to flip-flops and proportional amplifiers.

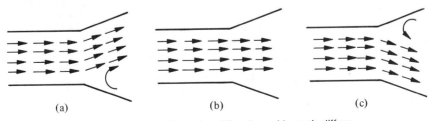

Figure 10.3. Possible modes of flow in a wide-angle diffuser.

In addition to the foregoing it has also been shown that a jet can have two or three stable states depending on the angle of the diffuser through which a jet is flowing. Flow can separate from one wall of a wide-angle diffuser and follow the other wall or stay between the two walls, as shown in Figure 10.3. For a certain angle of the diffuser walls there is a zone where flow can separate from one wall or from both and is stable in any of these states.

10.3 Fluid Amplifiers

When a fluidic device is operated as an amplifier it provides a means by which high-energy flows can be controlled by low-energy control signals. The control signals can be derived from any convenient source, such as compressed gas, high-pressure liquids, etc. Let us now consider the device shown in Figure 10.4. In this device a control jet is applied between the power jet and the wall to increase the pressure at that point and switch the jet to the opposite wall. The central splitter is used to obtain two separate outputs.

In Figure 10.4 the left-hand control signal has been used to obtain an output to the right receiver. If switching to the left is desired, the left control signal is removed and the right control signal is actuated to break the Coanda effect. It has been possible to obtain switching speeds of several hundred cycles per second with this configuration and it has been aptly called in the jargon of electronics a bistable flip-flop. This device is capable of performing a number of digital logic functions. In the flip-flop shown in Figure 10.4 the power jet will remain stable in either position due to the Coanda effect and will shift from one position to the other only if it is disturbed either by a pressure pulse or by continuous pressure from a control port. Obviously this device can also be used as a simple switch.

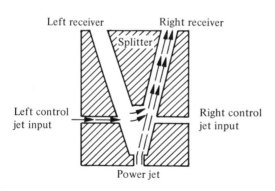

Figure 10.4. Fluid amplifier.

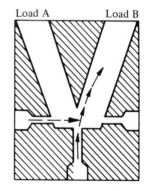

Figure 10.5. Fluidic amplifier.

By simply changing the geometry of the device shown in Figure 10.4 it is possible to design a device in which wall attachment cannot take place. The device shown in Figure 10.5 depends on the direct interation between the control jets and power jets to deflect the power stream. This device can readily be analyzed mathematically by considering the conservation of momentum in the system. The momentum entering this system in the power jet is

$$M_p = \dot{m}_p V_p \qquad (10.1)$$

where M_p is the momentum influx of the power jet, $\dot{m}_p$ the mass rate of flow of the power jet, and V_p the velocity of the power jet. Note that the momentum influx of the power jet is a vector directed vertically into the system based upon the orientation shown in Figure 10.5. The momentum of the control jet is

$$M_c = \dot{m}_c V_c \qquad (10.2)$$

where the subscript c indicates control jet and the other symbols are as for the jet. The momentum of the control jet is a vector to the right, as shown in Figure 10.5. Since the principle of conservation of momentum requires momentum to be conserved and momentum is a vector, we obtain the diagram shown in Figure 10.6. From this figure we can write directly

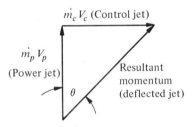

Figure 10.6. Conservation of momentum.

$$\tan \theta = \frac{\dot{m}_c V_c}{\dot{m}_p V_p} \qquad (10.3)$$

However,

$$\dot{m} = \rho A V \qquad (10.4a)$$

Therefore,

$$\tan \theta = \frac{(\rho a V^2)_c}{(\rho A V^2)_p} \qquad (10.4b)$$

In general the velocity leaving a jet can be obtained from the Bernoulli equation or Torricelli's equation, and it is found that

$$V^2 = KP \qquad (10.5)$$

where K is a constant and P the pressure of the jet. Substitution of (10.5) into (10.4) yields

$$\tan \theta = \frac{K_c P_c \rho_c}{K_p P_p \rho_p} \tag{10.6}$$

and if $\rho_c = \rho_p$,

$$\tan \theta = K' \frac{P_c}{P_p} \tag{10.7}$$

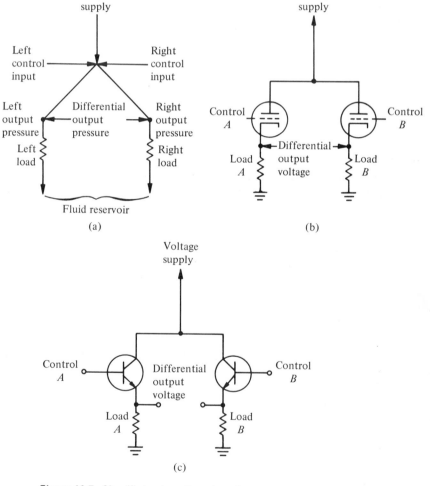

(a)

(b)

(c)

Figure 10.7. Simplified schematics of analogous amplifiers: (a) fluid, (b) vacuum tube, and (c) transistor.

From equation (10.7) it is apparent that the angle of deflection of the power jet is that angle whose tangent is proportional to the ratio of the control pressure to the power pressure. Based upon this conclusion this type of device has been named a proportional amplifier since the deflection of the power jet is proportional to the strength of the control signal.

In the device shown in Figure 10.5 the splitter divides the power jet evenly when a control jet is not present so that the differential pressure across the load is zero. When the power jet is deflected in either direction by a control jet input, a differential pressure occurs across the load since kinetic energy is reconverted to pressure energy by properly designing the passages in the device. When operated in this manner this device is essentially an analog unit. If gain is defined as the ratio of output change to control change, a gain of from 25 to 100 can be expected from this type of amplifier.

It is interesting to note the similarity (functionally) between the fluid amplifiers and electrical amplifiers. Figure 10.7 shows these analogous units with the fluid amplifiers controlled by a differential pressure signal.

Another form of amplifier is the turbulence amplifier, which depends on the change in flow conditions that accompany a change from laminar to turbulent flow in a fluid stream. When a laminar flowing fluid leaves a tube, it remains laminar for distances as great as 100 times the tube diameter. The point at which the flow becomes turbulent depends on the velocity of the stream or the presence of an external disturbance. Figure 10.8 shows the usual construction of a turbulence amplifier. A collector tube placed in the path of the stream measures output in terms of static pressure. A control jet is placed at right angles to the power jet. Small disturbances, created by control pressure, cause the stream's point of turbulence to shift drastically toward the supply tube, with a consequent sudden fall in output pressure. These devices are so sensitive that gains of 1000, or more, have been achieved, albeit at the cost of

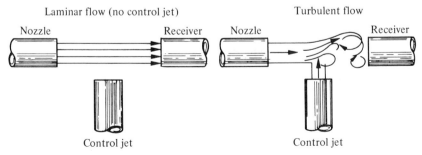

Laminar flow (no control jet)   Turbulent flow

Nozzle   Receiver   Nozzle   Receiver

Control jet   Control jet

Figure 10.8. Turbulence amplifier. Reproduced with permission from Introduction to Fluid Mechanics *by R. W. Henke, Addison-Wesley Publishing Company, Inc., Reading, Mass., 1966, p. 135.*

stability. They can be triggered acoustically, or even thermally, rather than by a control jet.

Just as the momentum exchange device described earlier can be used as a bistable flip-flop or fluid amplifier, the turbulence amplifier can be used in digital devices to perform both logic and switching functions. As a switch, use is made of the fact that the output pressure must be above some predetermined

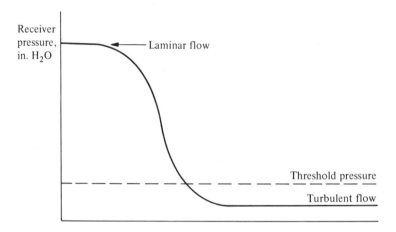

Figure 10.9. *Turbulence amplifier as a switch. Reproduced with permission from* Introduction to Fluid Mechanics *by R. W. Henke, Addison-Wesley Publishing Company, Inc., Reading, Mass., 1966, p.135.*

threshold pressure for the device to function. Referring to Figure 10.9, the minimum operating pressure below which the system will not operate is set above the lowest receiver pressure (which occurs during turbulent flow), providing on-off switching of the system.

The last category of amplifiers that will be considered in this chapter is the vortex amplifier, shown schematically in Figure 10.10. This device operates on the principle of momentum conservation by changing the path of the power jet by the control jet. As shown in Figure 10.10, this device consists of a hollow cylinder with a central outlet hole. If the control jet is not present, the flow from the power jet flows freely across the chamber to the outlet with very little impedance to the flow. Application of the control jet causes the power jet to rotate spirally due to the tangential application of the control flow and the resultant momentum transfer. As a result of this interaction a vortex (or whirlpool) is generated. The power jet now follows a path that causes it to spiral inward. To conserve angular momentum, the tangential velocity of the power jet is increased as it approaches the central outlet, which increases the shear stresses in the fluid, leading to a high-pressure drop in the power jet from the

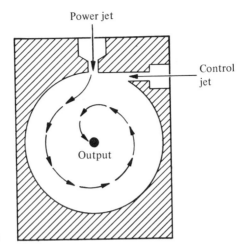

Power jet

Control
jet

Output

Figure 10.10. Vortex amplifier.

inlet to the outlet of the device. This high-pressure drop decreases the flow from the power jet, and in effect the device acts as a variable restrictor to decrease the power flow rather than to divert it.

10.4 Digital Fluidic Devices

Most fundamental switching and logic functions that can be performed electronically can also be performed using fluidic devices. The devices described in Section 10.3 are usually operated as analog devices, but with little or no modifications they can also be operated as digital devices. When describing some of the following digital applications of fluidic devices, a short discussion of logic functions as used in the mathematics of logic will be considered.

The OR gate will give a positive output in a device when input A or input B is positive. In terms of Boolean algebra this function is written as,

$$F = A + B \qquad (10.8)$$

where the implied addition $(+)$ is read as OR. Equation (10.8) is read as follows: Output (F) is positive when inputs A or B are positive. Figure 10.11 shows a fluidic OR gate in which an output is obtained if either control jet A or B is present. Due to the configuration, there is very little leakage between the two inputs. However, if a high impedance is present at the output, bleeds are provided to permit the fluid to escape before reversing and flowing into the wrong input. This device does not require a power source and is passive.

The NOR gate consists of an OR gate followed by an inverter stage where

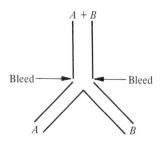

Figure 10.11. *Fluid OR gate with bleeds to prevent leakage.*

the inverter output is the inverse, or NOT, of the inverter input. This circuit provides a NOT-OR or NOR function. The Boolean expression for this gate is

$$F = \overline{A + B} \qquad (10.9)$$

where once again the plus symbol is read to be OR and the bar indicates NOT. Thus one reads equation (10.9) as follows: Output (F) equals NOT A OR B. The NOR fluidic gate shown in Figure 10.12a is obtained by combining an inverter (Figure 10.12b) with the OR gate of Figure 10.11. The inverter shown in Figure 10.12b, when operated in the absence of the control jet, allows the power jet to flow unimpeded in a straight path and the output indicates the absence of control, or $\bar{A}$. When the control is present, the power jet is switched to the receiver on the right and indicates the presence of control, or A. The Coanda effect in the right receiver is minimized, and the power jet switches back to the left receiver immediately after the control is removed.

In Boolean algebra the inverter equation is

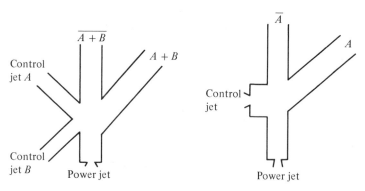

Figure 10.12. *(a) NOR gate; (b) pure-fluid inverter.*

$$F = \bar{A} \qquad (10.10)$$

which reads as follows: Output (F) equals NOT input A and the inverter is therefore a NOT gate.

In the NOR gate if neither control jet A nor B is present, the power jet flows straight through to the left receiver, indicating AB. However, the presence of A or B forces the power jet to switch to the right receiver, indicating $A + B$. Here again the Coanda effect is minimized in the right receiver to enable the power jet to switch back to the left receiver immediately when there is no control jet.

The AND gate has a positive output only when A AND B are positive. In terms of Boolean algebra this logic function can be written as

$$F = AB \qquad (10.11)$$

which can be interpreted as follows: Output (F) equals A and B. The fluidic AND gate is shown in Figure 10.13. This device is passive and does not require a power jet. If only control jet A or B is present, it passes straight through to one of the leakage paths. If both are present and their magnitudes are approximately equal, the combined jet makes a 45 deg turn with respect to the original jets in accordance with the momentum-transfer principle. The result is that the combined jet flows into the receiver to indicate AB.

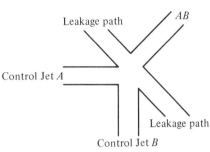

Figure 10.13. AND gate.

In the foregoing discussion some of the elements that go into a digital circuit have been briefly discussed. Many other basic digital elements exist, and the student is referred to the technical literature for a further discussion of digital components, Boolean algebra, and the mathematics of logic. However, some of the items noted thus far in this section can be summarized and made more evident if we consider a recent commercial device that utilizes some of these elements. Figure 10.14 shows the PB Flowboard, a fluidic unit containing 22 amplifiers, each of which is a pure-fluid device performing the NOR logic functions. Table 10.1 shows the four logic functions discussed in this section. The truth table shows the combinations of A and B that will or will not yield F, depending on the logic functions in question. The next item in this table is the conventional symbolism for each of the logic functions, and the following

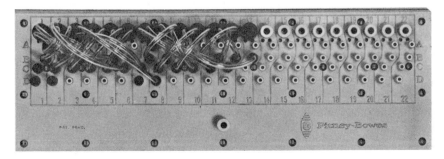

Figure 10.14. PB Flowboard. Courtesy of the Pitney-Bowes Corp., Stanford, Conn. The name PB Flowboard is a trade name of the Pitney-Bowes Corp.

column is the relay equivalent (electrical) symbolism for the corresponding logic function.

The last entry in Table 10.1 shows the connections required on the PB Flowboard to achieve the desired logic function. It must be borne in mind that each of the 22 amplifiers is basically capable of performing only the NOR logic function and that the connections shown are required to achieve any of the other functions. While relatively simple in principle, the complexity of the interconnections necessary for this unit can be judged from Figure 10.14. Standard connectors, plugs, and accessories have been developed for this unit and are available commercially. Some of the advantages claimed for this unit are common manifold requiring only one fluid supply connection, a single compact structure containing 22 logic elements, low-pressure consumption, and reliable operation during vibration.

Table 10.1 Basic Logic Functions and Connections[†]

LOGIC FUNCTION	TRUTH TABLE			SYMBOLISM	RELAY EQUIVALENT	AMPLIFIER CONNECTIONS
NOR	A	B	F	A— $\rceil$1 $\rangle$o—F, B— $F = \overline{A+B}$	A B, —N N—F	
	0	0	1			
	1	0	0			
	0	1	0			
	1	1	0			
NOT	A		F	A— $\rceil$1 $\rangle$o—F, $F = \overline{A}$	A, —N—F	
	1		0			
	0		1			
OR	A	B	F	A— $\rceil$1 $\rangle$ $\rceil$2 $\rangle$o—F, B—, $F = A + B$	A, —H—, —H—F, B	
	0	0	0			
	1	0	1			
	0	1	1			
	1	1	1			
AND	A	B	F	A— $\rceil$1 $\rangle$ $\rceil$3 $\rangle$o—F, B— $\rceil$2 $\rangle$, $F = A \cdot B$	A B, —H H—F	
	0	0	0			
	1	0	0			
	0	1	0			
	1	1	1			

[†]Provided courtesy of the Pitney-Bowes Corp., Stanford, Conn. The name PB Flowboard is a trade name of the Pitney-Bowes Corp.

10.5 Future of Fluidics

To understand the rapidity of the advance and to gauge the possible future of fluidics, it is necessary to take a brief look at the applications developed to date for these devices. They are being tried in a host of industries.

In the machine tool industry, fluidics, primary limitiation—relatively low-speed operation compared to the electronic devices—is not important. Switching speeds of digital fluidic elements are approximately 1 msec, sufficiently fast for most machine tools. Reliability and simplicity appear to be more important than speed in attracting machine tool builders to fluidics for control purposes. In one instance a machine tool manufacturer converted to fluidic control on a battery of 16 of its own milling machines, cutting flutes on bar stock to make twist drills; another introduced fluidic controllers on grinding machines, a third on turret lathes, and still another on door-punching machines.

Various specialty machinery makers and users have also begun to incorporate fluidic devices into their equipment. One manufacturer designed a very elaborate system for the company's glass-making press. The entire press operates on air with only one electric drive motor; it is completely automatic, from cutting the semimelted glass to final shaping and trimming of the end product. An earlier application, apparently quite successful, used fluid logic to control automatic machinery that makes expansion watch bracelets.

A great deal of work has been done to find suitable applications for fluidics in the aerospace industry. Although response speed is a severe limitation, high reliability is important in military equipment that operates in diverse and hostile environments. Recently two manufacturers announced the development of fluidic fuel control systems for commercial turbine engines. Automatic flight controls have also been developed in which all elements of the autopilot are fluidic. Test results from an Aero Commander 680FP airplane are comparable with conventional systems. A fluidic stability augmentation system has been devised to replace existing electronic equipment of the Boeing CH46A helicopter, and it has also been reported that stabilization of the Bell UH-1c helicopter has been achieved with fluidics. Initial results indicate that an MTBF (Mean Time Between Failure) of 38,000 hr is obtainable for this system, compared with only 9000 hr for electromechanical systems.

Not many automobile applications have been reported to date. The automobile industry generally does not use equipment unless it has been successfully demonstrated over a long period. Nevertheless, an interesting application has been announced quite recently—a fluidic carburetor. The unit is exceedingly simple; it has only one moving part, the throttle. It is compact, it is cheap when manufactured in quantity, and it shows good response and even increased fuel economy by avoiding wastage in the accelerator pump and evaporation from the float bowl.

A significant achievement has been the introduction of fluidic controls to electrical drive motors in new locomotives, where they replace electromechanical systems. Over-the-road tests with prototype devices have been quite successful.

10.6 Closure

Fluidic systems have established themselves as an integral part of modern technology. While fluidic systems will not replace electronic systems, they will be suited for use in environments that are basically hostile to electronic systems such as regions of high nuclear radiation, high temperature, and intense shock and vibration. In addition, fluidics will also be found in those special situations where simplicity and/or reliability justifies the replacement of a complex electronic circuit. One area of promise that has already materialized is the area of process and machine control. Fluidic systems and devices have successfully proved themselves in this area, and it is anticipated that further utilization of these devices will occur in this field.

Modern fluidics dates from approximately 1960 and is a relative newcomer to the technological scene. Due to its relative newness, its promise, and its possible application to many different technical areas, it is a branch of technology that has generated a great deal of interest by the technical community. It is to be expected that changes in this field will occur very rapidly, and a special caution is advanced to the student to be aware of these changes and to be familiar with current technical publications if he is to work in this field.

Appendix **A**

ANSWERS TO PROBLEMS

Chapter 1

1.1 68 ; 104 ; 140
1.2 −17.7 ; −12.2 ; 10
1.3 4.44 ; 227 ; 327 ; 87.8
1.4 °C = 20/11 (°ARB-20)
1.5 −130.15° ARB
1.6 Derivation
1.7 Derivation
1.8 39.5 psia ; 7 ft
1.9 39.5 psia ; 6.7 ft
1.10 10.3 psia
1.11 0.87 psia
1.12 0.78 ; 0.04

1.13 40
1.14 Derivation
1.15 37.6 ft/sec
1.16 −0.104 ft
1.17 −0.052 ft
1.18 +1 × 10^6 lb/in².

Chapter 2

2.1 40.7 psia
2.2 427 psig
2.3 66.1 lb/ft³
2.4 19,700 ft
2.5 Infinity

2.6   No, by 102 ft

2.7   2578 psfa

2.8   10 psi

2.9   0.0735 psi

2.10 −4.75 psf

2.11 18.72 psia

2.12 1.14 psf

2.13 16 in.

2.14 17.8 psia

2.15 14.75 in.

2.16 0.433 psi

2.17 $\frac{2}{3} \gamma R^3$

2.18 5200 lb

2.19 70,800 lb ; 14.5 ft

2.20 $F = \frac{1}{2}\gamma h^2(h \cot \theta + b)$

2.21 12,000 lb

2.22 $h_{cp} = \dfrac{d^2(B^2 + 4Bb + b^2)}{12(2B + b)[\frac{1}{2}(B + b) + b]}$
      $+ \, d\left(\dfrac{2B + b}{3(B + b)}\right);$
      $B = 2h \cot \theta + b$

2.23 3.13 ft from bottom

2.24 156 lb/ft³

2.25 258 lb

2.26 643,000 lb

2.27 33.6 lb ; $x = 4.24$ in. ; $y = 6.67$
      in. ; $\theta = 57.5$ deg

2.28 $F = h^2 \gamma_W [(4 + W)/3]$

2.29 0.12 in.

2.30 0.06 in.

2.31 312 psi

Chapter 3

3.1 2.25×10^6 lb/hr

3.2   2.04 ft/sec

3.3   Derivation

3.4   Derivation

3.5 $V_{10} = \frac{8}{25} V_4$

3.6   348 gal/min

3.7   1000 ft lb

3.8   80.3 ft/sec

3.9   388 ft lb ; 38.8 ft

3.10 197 ft/sec ; 600 ft lb/lb

3.11 113.5 hp

3.12 114.9 ft/sec ; 80.3 ft

3.13 557 hp

3.14 834.5 ft lb

3.15 −2160 ft lb/lb

3.16 208° F

3.17 Zero ; neither

3.18 2330 ft/sec

3.19 2.57 lb

3.20 1.33

Chapter 4

4.1   9.72 ft

4.2   7.67 ft/sec ; 0.914 ft ; 91.7 ft ;
      102.6 ft

4.3   58.6 gal/min

4.4   98.3 ft/sec

4.5   8520 hp

4.6   2730 gal/min ; 31 ft/sec

4.7   1585 gal/min ; 29.4 ft/sec

4.8   16.17 psia

4.9   83.3 percent

4.10 92.2 percent ; 2250 lb

4.11 2700 lb

4.12 5570 hp

4.13 95.4 percent

4.14 1.24 ft³/sec

4.15 5.16 ft/sec

4.16 $y = x^2/4h$

4.17 8.01 ft/sec

4.18 $C = 0.9$

4.19 22.2 ft/sec

4.20 4.36 psi

4.21 66.1 ft/sec

4.22 3.74 psi

4.23 0.29 in.

4.24 0.357 in.

4.25 Yes

4.26 Derivation

4.27 Yes

4.28 340 lb

4.29 152 lb
4.30 63.4 lb
4.31 8.05 hp
4.32 9.6 hp

Chapter 5

5.1 Identical to M-L-T-θ solution
5.2 $f = $ Constant $(kg/W)^{1/2}$
5.3 $p/\rho V^2 = $ Constant $(DV\rho/\mu)^{-b}$
     $\times (V^2/gD)^{-f} (e/D)^e (L/D)^g$
5.4 $F/\rho D^2 V^2 = $ Constant $(DV\rho/\mu)^{-d}$
5.5 $f = $ Constant $(k/M)^{1/2}$ or
     $f = $ Constant $(g/L)^{1/2}$
5.6 $p = $ Constant (γh)
5.7 $V_s = $ Constant $\sqrt{pg/\gamma}$
5.8 $V = $ Constant $(D^2\gamma/\mu)$
5.9 $\delta/L = $ Constant $(F/EL^2)^a (I/L^4)^d$
5.10 $P/D^2 pw = $ Constant $(D^2 \rho w^2/p)^b$
     $\times (g/Dw^2)^e$
5.11 50,000
5.12 $V_2/V_1 = 6$
5.13 $V_p = \frac{1}{12}$ ft/sec
5.14 Force, length, velocity, density,
     viscosity
5.15 $V_p = 3.46$ ft/sec
5.16 $V_M = 450$ mph
5.17 $V_8 = 112.5$ ft/sec

Chapter 6

6.1 Derivation
6.2 62.7

6.3 $D_{\text{eq}} = 4\left[\dfrac{\pi D^2}{4}\left(1 - \dfrac{\cos^{-1}\dfrac{h}{D}}{\dfrac{2}{180}}\right)\right.$

$\left. + \left(\sqrt{\left(\dfrac{D}{2}\right)^2 - h}\right)\Big/h\right]$

$\times \dfrac{180}{\pi D}\cos^{-1}\dfrac{h}{\dfrac{D}{2}}$

6.4 0.55 ft/sec
6.5 0.055 ft/sec
6.6 3.24×10^{-4} ft
6.7 6.27×10^{-5} lb/ft^2
6.8 248 ft/sec
6.9 $D_{eq} = D_2 - D_1$
6.10 9.64 psi/ft
6.11 Turbulent
6.12 636 gal/min
6.13 103.5 spig
6.14 28.2 psi
6.15 0.0136 psi/ft
6.16 74 psi
6.17 75.5 psi
6.18 156.6 psig
6.19 390 psi
6.20 118.6 ft of water
6.21 36.5 psi
6.22 0.61 ft
6.23 0.174 ft
6.24 81.4 ft
6.25 109 ft
6.26 0.58 ft
6.27 14 ft
6.28 1 psi
6.29 1.4 psi
6.30 10.4 psi
6.31 75.6 psi

Chapter 7

7.1 $M = 0.09$
7.2 $h = 120.8$ Btu/lb ; $T = 504°$ R
7.3 $T = 320.8°$ R ; $p = 27.5$ psia
7.4 $435°$ R
7.5 52.8 psia ; 1060 ft/sec
7.6 1.22
7.7 360 ft/sec
7.8 $5.05°$ F
7.9 0.27 lb/sec
7.10 91.4 percent
7.11 Derivation

7.12 Derivation

7.13 As assigned

Chapter 8

8.1 (a) $4Wh/(2h + W)$

(b) $\left(\dfrac{\sqrt{3}}{2}\right)S$

(c) $4W/3$

8.2 (a) $2Wh/(W + h)$

(b) $\left(\dfrac{\sqrt{3}}{4}\right)S$

(c) W

8.3 313 ft³/sec

8.4 267 ft³/sec

8.5 1.885

8.6 204 ft³/sec

8.7 1.55 ft

8.8 1 : 66 ; 200

8.9 1 : 56.2

8.10 17.3 ft wide ; 8.65 ft high

8.11 4.05 ft wide ; 2.02 ft high

8.12 4.5 ft wide ; 2.25 ft high

8.13 4.24 ft

8.14 Tranquil ; 14.88 ft

8.15 0.36 ft ; 4.96 ft

8.16 70 ft³/sec

8.17 7.86 ft

8.18 3.22 ft

8.19 57.6 ft³/sec

8.20 2.3 ft

8.21 2.31 ft

8.22 65.2 ft

Chapter 9

9.1 1.7×10^{-3} in.

9.2 2.34×10^{-10} lb

9.3 $C_D = 34/\text{Re}$

9.4 $D = 4.25 \times 10^{-7}$ lb

9.5 1410 lb

9.6 0.0230 lb

9.7 0.00185 hp

9.8 0.216 lb

9.9 2.72

9.10 $L = 3.27$ lb ; $D = 0.717$ lb ; $R = 3.34$ lb

9.11 582 lb

9.12 175 lb

9.13 $L = 33,500$ lb ; $D = 2,250$ lb ; hp $= 1,080$ lb ; $L/D = 14.9$

9.14 4.5 deg

9.15 69.4 ft²

9.16 66.8 lb/ft²

9.17 33.4 lb/ft²

9.18 158 mph

9.19 0.894

9.20 0.894

9.21 (a) $V = 127$ ft/sec ; $D = 655$ lb; hp $= 151$

(b) $V = 139$ ft/sec ; $D = 526$ lb; hp $= 133$

9.22 3 deg

9.23 $L = 8900$ lb

9.24 $C_{D_i} = 0.0135$ deg

9.25 $\alpha_i = 0.97$ deg

9.26 $C_{D_9} = 0.021$; $\alpha_9 = 1.49$

9.27 73.5 percent

9.28 $C_D = 0.083$

9.29 $\alpha_8 = 4$ deg ; C_{D_8} 0.032

9.30 $\theta = 2.7$ deg

9.31 $V_T = 684$ ft/sec

9.32 $V_T = 914$ ft/sec

9.33 3.06 miles

9.34 $\theta = 36.1$ deg

9.35 g load $= 1.41$; $R = 2170$ ft

9.36 g load $= 3.86$

9.37 R.C. $= 27.5$ ft/sec

PRESSURE
MEASUREMENTS[1]

B.1 Introduction

Pressure is a property of a system and as a fundamental parameter it is of the utmost importance that it be measured accurately. In Chapter 1 pressure was defined to be the normal force exerted by a fluid on a surface. It would be more correct to qualify this definition and for the present restrict it to fluids at rest. However, no definition is really useful unless and until it can be converted into measurable characteristics. Figure B.1 illustrates the basic definition of pressure and the elementary static concepts upon which the entire field of pressure measurement is based.

There is unfortunately a confusing number of units used to express pressure. While all of these units are based upon the definition of a normal force per unit area, their usage has entered the technical literature based upon particular applications. Table B.1 gives conversion factors between some of the commonly

[1]The material in this appendix is taken with permission from "Pressure and Its Measurement" by R. P. Benedict, *Electro-Technology (New York)* **80**, October 1967, p. 69.

Table B.1 Pressure-Unit Conversion Factors

Pressure Unit	psi	in. H_2O	in. Hg	Atmos- pheres	Microbars	mm Hg	Microns
1 psi	1.000	27.730	2.0360	6.8046 $\times 10^{-2}$	68,947.6	51.715	51,715.0
1 in. H_2O (68°F)	0.036063	1.000	0.073424	2.4539 $\times 10^{-3}$	2,486.4	1.8650	1,865.0
1 in. Hg (32°F)	0.49115	13.619	1.000	3.3421 $\times 10^{-2}$	33,864.0	25.400	25,400.0
1 atm	14.69595	407.513	29.9213	1.000	1.01325 $\times 10^6$	760.000	7.6000 $\times 10^5$
1 microbar (dynes /cm²)	1.4504 $\times 10^{-5}$	4.0218 $\times 10^{-4}$	2.9530 $\times 10^{-5}$	9.8692 $\times 10^{-7}$	1.000	7.5006 $\times 10^{-4}$	0.75006
1 mm Hg (32°F)	0.019337	0.53620	0.03937	1.3158 $\times 10^{-3}$	1,333.2	1.000	1,000.0
1 micron (32°F)	1.9337 $\times 10^{-5}$	5.3620 $\times 10^{-4}$	3.9370 $\times 10^{-5}$	1.3158 $\times 10^{-6}$	1.3332	0.0010	1.000

used units of pressure. For convenience, there are two approximations that can be used to give the student a "feel" for some of these units, namely

$$1 \text{ mm} = 1000 \text{ microns} \simeq 0.04 \text{ in. Hg} \quad \text{and} \quad 1 \text{ micron} = 0.001 \text{ mm} \simeq 0.00004 \text{ in. Hg}$$

Table B.2 Characteristics of Pressure Standards

Type	Range	Accuracy
Dead-weight piston gage	0.01 to 10,000 psig	0.01 to 0.05 percent of reading
Manometer	0.1 to 100 psig	0.02 to 0.2 percent of reading
Micro- manometer	0.0002 to 20 in. H_2O	1 percent of reading to 0.001 in. H_2O
Barometer	27 to 31 in. Hg	0.001 to 0.03 percent of reading
McLeod gage	0.01 micron to 1 mm Hg	0.5 to 3 percent of reading

In performing any measurement it is necessary to have a standard of comparison in order to calibrate the measuring instrument. In Section B.2 five

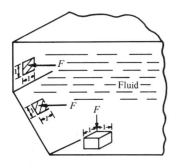

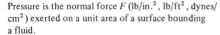

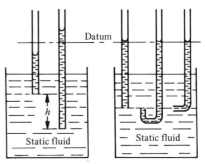

Pressure is the normal force F (lb/in.², lb/ft², dynes/ cm²) exerted on a unit area of a surface bounding a fluid.

Fluid pressure varies with depth, but it is the same in all directions at a given depth.

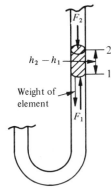

Variation of fluid pressure with elevation is found by balancing the forces on a static-fluid element (F_1 is equal to F_2 plus the weight of the element). For a constant-density fluid, the pressure difference $p_2 - p_1$ is equal to the specific weight γ times $(h_2 - h_1)$.

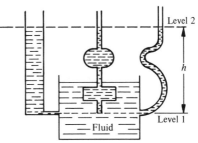

Pressure is independent of the shape and size of the vessel. The pressure difference between level 1 and level 2 is always $p_1 - p_2 = \gamma h$, where γ is the specific weight of the constant-density fluid in the vessel.

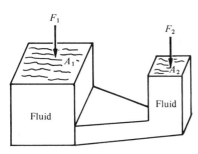

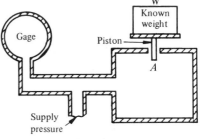

Constant pressure transmission in a confined fluid can be used to multiply force by the relation $p = F_1/A_1 = F_2/A_2$.

When the known weight is balanced, the gage pressure is $p = W/A$; this is the basic principle of dead-weight testing.

Figure B.1. *Basic pressure-measurement concepts.*

pressure standards currently used as the basis for all pressure-measurement work will be discussed. Table B.2 summarizes these standards, the pressure range in which they are used, and their accuracy.

B.2 Dead-Weight Piston Gage

The dead-weight, free-piston gage consists of an accurately machined piston inserted into a close-fitting cylinder. Masses of known weight are loaded on one end of the free piston, and pressure is applied to the other end until enough force is developed to lift the piston-weight combination. When the piston is floating freely between the cylinder limit stops, the gage is in equilibrium with the unknown system pressure. The dead-weight pressure can then be defined as

$$P_{dw} = \frac{F_e}{A_e} \tag{B.1}$$

where F_e, the equivalent force of the piston-weight combination, depends on such factors as local gravity and air buoyancy, and A_e, the equivalent area of the piston-cylinder combination, depends on such factors as piston-cylinder clearance, pressure level, and temperature.

A fluid film provides the necessary lubrication between the piston and cylinder. In addition, the piston—or, less frequently, the cylinder—may be rotated or oscillated to reduce friction even further. Because of fluid leakage, system pressure must be continuously trimmed upward to keep the piston-weight combination floating. This is often achieved by decreasing system volume using a pressure volume apparatus (as shown in Figure B.2). As long as the piston is freely balanced, system pressure is defined by equation (B.1).

Corrections must be applied to the indication of the dead-weight piston gage, p_i, to obtain the accurate system pressure, P_{dw}. The two most important corrections concern air buoyancy and local gravity. According to Archimedes' principle, the air displaced by the weights and the piston exerts a buoyant force

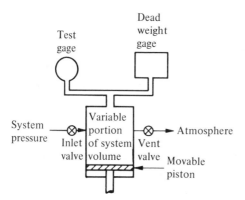

Figure B.2. Pressure volume regulator to compensate for fluid leakage in a dead-weight gage.

that causes the gage to indicate too high a pressure. The correction term for this effect is the ratio of two specific weights,

$$C_b = -\frac{\gamma_{air}}{\gamma_{weights}}$$

Whenever gravity at the measurement site varies from the standard sea-level value of 32.1740 ft/sec² because of latitude or altitude variations, a gravity correction term must be applied;

$$C_g = \frac{g_{local}}{g_{std}} - 1$$
$$= -2.637 \times 10^{-3} \cos 2\phi - 9.6 \times 10^{-8}h - 5 \times 10^{-5} \qquad (B.2)$$

where ϕ is latitude in degrees and h is altitude above sea level in feet.

The corrected dead-weight piston gage pressure is therefore

$$P_{dw} = p_i(1 + C_b + C_g) \qquad (B.3)$$

The effective area of the dead-weight piston gage is usually taken as the mean of the cylinder and piston areas, but temperature affects this dimension. The effective area increases between 13 and 18 ppm/°F for commonly used materials, and a suitable correction for this effect may also be applied.

B.3 Manometer

The manometer has been discussed earlier in this book as a means of measuring pressure. However, for accurate measurements there are certain factors that must be taken into account. Therefore, for the sake of continuity and completeness a brief discussion of the manometer will be repeated in this section, and those corrections most necessary for accurate use of this instrument will also be discussed.

A manometer can simply consist of a transparent tube in the form of an elongated U, partially filled with a suitable liquid. Mercury and water are the most commonly used manometric fluids because detailed information is available on their specific weights.

The fluid whose pressure is unknown is applied to the top of one of the tubes of the manometer, while a reference fluid pressure is applied to the other tube, as shown in Figure B.3. In the steady state, the difference between the unknown pressure and the reference pressure is balanced by the weight/unit area of the displaced manometer liquid, so that

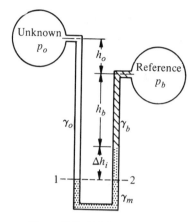

Figure B.3. *U-tube manometer.*

$$\Delta p_{\text{mano}} = \gamma_m \, \Delta h_e \qquad (B.4)$$

where γ_m, the corrected specific weight of the manometer fluid, depends on such factors as temperature and local gravity, and h_e, the equivalent manometer fluid height, depends on such factors as scale variations with temperature, relative specific weights, heights of the fluids involved, and capillary effects. This equation holds as long as manometer fluid displacement is constant.

The value of specific weight γ_m corrected for local gravity is

$$\gamma_c = \gamma_{\text{s.t}}(1 + C_g) \qquad (B.5)$$

where C_g is the correction term given in equation (B.3); γ_c is the corrected value of specific weight for any fluid whose specific weight is $\gamma_{\text{s.t}}$ at standard gravity (32.1740 ft/sec^2) and a given Fahrenheit temperature t.

The variation of specific weight with temperature is described by empirical relationships; for mercury,

$$\gamma_{\text{s.t}} = \frac{0.491154}{[1 + 1.01(t - 32) \times 10^{-4}]} \qquad (B.6)$$

and for water,

$$\gamma_{\text{s.t}} = \frac{(62.2523) + 0.978476 \times 10^{-2}t - 0.145 \times 10^{-3}t^2 + 0.217 \times 10^{-6}t^3)}{1728}$$

$$(B.7)$$

where $\gamma_{\text{s.t}}$ is in pounds per cubic inch. These relationships have proved satisfactory for accurate manometeter measurements; some typically useful values are listed in Table B.3.

One important correction needed to find the equivalent manometer fluid height h_e is associated with the relative weights and heights of the fluids involved. This hydraulic correction is

$$C_h = \left\{ 1 + \left(\frac{\gamma_b}{\gamma_m} \right)\left(\frac{h_b}{\Delta h_i} \right) - \left(\frac{\gamma_a}{\gamma_m} \right)\left[\left(\frac{h_a + h_b}{\Delta h_i} \right) + 1 \right] \right\} \qquad (B.8)$$

The quantities in this equation are depicted in Figure B.3.

Table B.3 Specific Weight of Mercury and Water
(At Standard Gravity Value of 32.1740 ft/sec²)

Temperature (°F)	Specific weight ($\gamma_{s,t}$)	
	Hg	H₂O
32	0.491154	0.036122
36	0.490956	0.036126
40	0.490757	0.036126
44	0.490559	0.036124
48	0.490362	0.036120
52	0.490164	0.036113
56	0.489966	0.036104
60	0.489769	0.036092
64	0.489572	0.036078
68	0.489375	0.036062
72	0.489178	0.036045
76	0.488981	0.036026
80	0.488784	0.036005
84	0.488588	0.035983
88	0.488392	0.035958
92	0.488196	0.035932
96	0.488000	0.035905
100	0.487804	0.035877

The second main correction needed to determine h_e concerns capillary effects. The shape of the interface between two fluids at rest depends on the relative gravity and cohesion and adhesion forces between the fluids and their containing walls. In water-air-glass combinations, the crescent shape of the liquid surface (called the meniscus) is concave upward, and the water is said to wet the glass. In this situation, adhesive forces dominate, and water in a tube will be elevated by capillary action. For mercury-air-glass combinations, cohesive forces dominate, the meniscus is concave downward, and the mercury level in a tube will be depressed by capillary action, as shown in Figure B.4.

Based upon the discussion of surface tension in Chapter 1, the capillary correction factor for manometers can be written as

$$C_c = \frac{2\cos\theta_m}{\gamma_m}\left(\frac{\sigma_{a-m}}{r_a} - \frac{\sigma_{b-m}}{r_b}\right) \qquad (B.9)$$

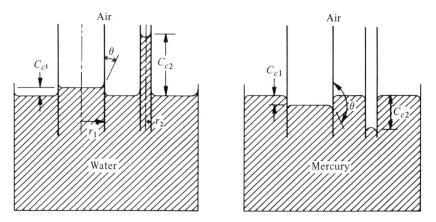

Figure B.4. *Meniscus effects in water and mercury.*

where θ_m is the angle of contact between the manometer fluid and the glass, σ_{a-m} and σ_{b-m} are the surface-tension coefficients of the manometer fluid (m) with respect to the fluids (a and b) above it, and r_a and r_b are the radii of the tubes containing fluids a and b. Typical values for these capillary variables are given in Table B.4. For mercury manometers, C_c is positive when the larger capillary effect occurs in the tube in which the manometer fluid is highest. For water manometers under the same conditions, C_c is negative.

Table B.4 Capillary Effects

Combination	Surface Tension σ		Contact Angle θ (deg)
	dynes/cm	lb/in.	
Mercury, vacuum, glass	480	2.74×10^{-3}	140
Mercury, air, glass	470	2.68×10^{-3}	140
Mercury, water, glass	380	2.17×10^{-3}	140
Water air, glass	73	0.416×10^{-3}	0

When the same fluid is applied to both legs of a standard U-tube manometer, the capillary effect is often neglected. This can be done because the tube bores are approximately equal, and capillarity in one tube counterbalances that in the other. However, capillary effect can be extremely important and must always be considered in manometer-type instruments.

To minimize the effect of a variable meniscus, which can be caused by dirt, the method of approaching equilibrium, tube bore, and so on, the tubes are always tapped before reading and the measured liquid height is always based on readings taken at the center of the meniscus in each leg of the manometer. To reduce the capillary effect itself, the use of large-bore tubes (over $\frac{3}{8}$-in. diameter) is most effective.

When both hydraulic and capillary corrections are taken into account, the equivalent manometer fluid height is

$$\Delta h_e = C_h \, \Delta h_i \pm C_c \qquad \text{(B.10)}$$

where Δh_i is the indicated height shown in Figure B.3.

B.4 Micromanometer

While the manometer is a useful and convenient tool for pressure measurements it is unfortunately limited when making low-pressure measurements. To extend its usefulness in the low-pressure range, micromanometers have been developed that have extended the useful range of low-pressure manometer measurements to pressures as low as 0.0002 in. H_2O.

One type of micromanometer is the so-called Prandtl-type micromanometer in which capillary and meniscus errors are minimized by returning the meniscus of the manometer liquid to a null position before measuring the applied pressure difference. As shown in Figure B.5, a reservoir, which forms one side of the

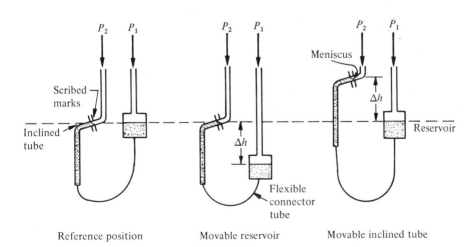

Figure B.5. *Two variations of the Prandtl-type micromanometer.*

manometer, is moved vertically to locate the null position. This position is reached when the meniscus falls within two closely scribed marks on the near-horizontal portion of the micromanometer tube. Either the reservoir or the inclined tube is then moved by a precision lead-screw arrangement to determine the micromanometer liquid displacement (Δh), which corresponds to the applied pressure difference. The Prandtl-type micromanometer is generally accepted as a pressure standard within a calibration uncertainty of 0.001 in. H_2O.

Another method for minimizing capillary and meniscus effects in manometry is to measury liquid displacements with micrometer heads fitted with adjustable, sharp index points. Figure B.6 shows a manometer of this type; the

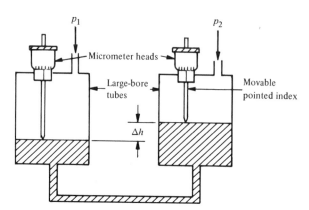

Figure B.6. *Micrometer-type manometer.*

micrometers are located in two connected transparent containers. In some commercial micromanometers, contact with the surface of the manometric liquid may be sensed visually by dimpling the surface with the index point, or even by electrical contact. Micrometer-type micromanometers also serve as pressure standards within a calibration uncertainty of 0.001 in. H_2O.

An extremely sensitive high-response micromanometer uses air as the working fluid and thus avoids all the capillary and meniscus effects usually encountered in liquid manometry. In this device, shown in Figure B.7, the reference pressure is mechanically amplified by centrifugal action in a rotating disk. The disk speed is adjusted until the amplified reference pressure just balances the unknown pressure. This null position is recognized by observing the lack of movement of minute oil droplets sprayed into a glass indicator tube. At balance, the air micromanometer yields the applied pressure difference

$$\Delta p_{\mathrm{micro}} = Kpn^2 \qquad (B.11)$$

where p is the reference air density, n the rotational speed of the disk, and K

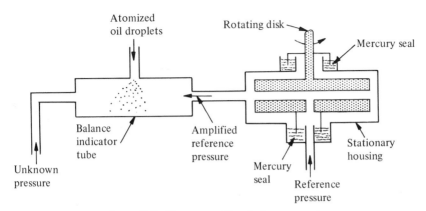

Figure B.7. Air-type centrifugal micromanometer.

a constant that depends on disk radius and annular clearance between the disk and the housing. Measurements of pressure differences as small as 0.0002 in. H_2O can be made with this type of micromanometer within an uncertainty of 1 percent.

A word on the reference pressures employed in manometry is pertinent at this point in the discussion. If atmospheric pressure is used as a reference, the manometer yields gage pressures. Because of the variability of air pressure, gage pressures vary with time, altitude, latitude, and temperature. If, however, a vacuum is used as reference, the manometer yields absolute pressures directly, and it may serve as a barometer. In any case, the absolute pressure is always equal to the sum of the gage and ambient pressures; by ambient pressure we mean the pressure surrounding the gage, which is usually atmospheric pressure.

B.5 Barometers

The reservoir or cistern barometer consists of a vacuum-reference mercury column immersed in a large-diameter ambient-vented mercury column that serves as a reservoir. The most common cistern barometer in general use is the Fortin type, in which the height of the mercury surface in the cistern can be adjusted. The operation of this instrument can best be explained with reference to Figure B.8.

The datum-adjusting screw is turned until the mercury in the cistern makes contact with the ivory index, at which point the mercury surface is aligned with zero on the instrument scale. Next, the indicated height of the mercury column in the glass tube is determined. The lower edge of a sighting ring is lined up with the top of the meniscus in the tube. A scale reading and a vernier reading

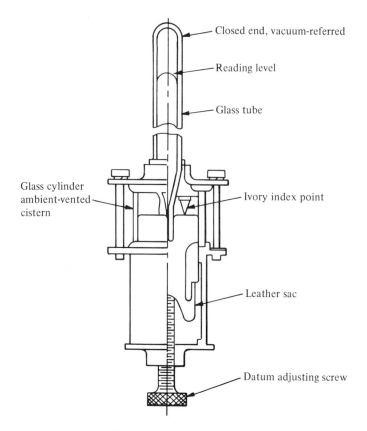

Closed end, vacuum-referred

Reading level

Glass tube

Glass cylinder
ambient-vented
cistern

Ivory index point

Leather sac

Datum adjusting screw

Figure B.8. Fortin barometer.

are taken and combined to yield the indicated mercury height at the barometer temperature.

Since atmospheric pressure on the mercury in the cistern is exactly balanced by the weight per unit area of the mercury column in the glass tube,

$$p_{\text{baro}} = \gamma_{\text{Hg}} h_{to} \tag{B.12}$$

The referenced specific weight of mercury, γ_{Hg}, depends on such factors as temperature and local gravity; the referenced height of mercury, h_{to}, depends on such factors as thermal expansion of the scale and the mercury.

When the scale zero of the Fortin barometer is adjusted to agree with the mercury level in the cistern, the correct height of mercury at temperature t will be

$$h_t = h_{ti}[1 + S(t - t_s)] \tag{B.13}$$

were S is the linear coefficient of thermal expansion of the scale, h_{ti} the indicated height of mercury, and t_s the temperature at which the scale was calibrated. Whenever t is greater than reference temperature t_o, the correct height of mercury at t will be greater than the height of mercury at t_o:

$$h_t = h_{to}[1 + m(t - t_o)] \tag{B.14}$$

where m is the cubical coefficient of thermal expansion of mercury and h_{to} the referenced height of the mercury column.

A temperature correction factor can be defined by

$$C_t = h_{to} - h_{ti} \tag{B.15}$$

From equation (B.13) and (B.14),

$$C_t = \left[\frac{S(t - t_s) - m(t - t_o)}{1 + m(t - t_o)} \right] h_{ti} \tag{B.16}$$

When standard values of $S = 10.2 \times 10^{-6}/°\text{F}$, $m = 101 \times 10^{-6}/°\text{F}$, $t_s = 62°\text{F}$, and $t_o = 32°\text{F}$ are substituted in equation (B.16),

$$C_t = -\left[\frac{9.08(t - 28.63)10^{-5}}{1 + 1.01(t - 32)10^{-4}} \right] h_{ti} \tag{B.17}$$

This temperature correction is zero at $t = 28.63°\text{F}$ for all values of h_{ti}. It can be approximated with very little loss of accuracy as

$$C_t \simeq -9(t - 28.6)10^{-5} h_{ti} \tag{B.18}$$

The uncertainty introduced in h_{to} by equation (B.18) is always less than 0.001 in. Hg for values of h_{ti} from 28.5 to 31.5 in. Hg and values of t from 60° to 100°F.

The specific weight γ_{Hg} in equation (B.12) must be based upon the local value of gravity and the reference temperature t_o; $\gamma_{Hg} = \gamma_{s,to}(1 + C_g)$, where C_g is the gravity correction term given by equation (B.2) and $\gamma_{s,to}$ is 0.491154 lb/in.³ Hg when $t_o = 32°\text{F}$.

When the barometer is read at an elevation other than that of the test site, an altitude correction factor must be applied to the local absolute barometric pressure. The altitude correction factor may be obtained from the pressure-height relationship derived in Chapter 2. When this is done we obtain

$$C_z = p_{\text{baro}} \left[e^{(Z_{\text{baro}} - Z_{\text{site}})/RT} - 1 \right] \tag{B.19}$$

where Z_{baro} and Z_{site} are altitudes in feet, R is the gas constant ($53.35\,\text{ft}/^\circ\text{R}$), and T is the absolute temperature in degrees Rankine.

Other factors may also contribute to the uncertainty of h_{ti}. Proper illumination is essential to define the location of the crown of the meniscus. Precision meniscus sighting under optimum viewing conditions can approach ± 0.001 in. With proper lighting, contact between the ivory index and the mercury surface in the cistern can be detected to much better than ± 0.001 in.

To keep the uncertainty in h_{ti} within 0.01 percent ($\simeq 0.003$ in. Hg), the mercury temperature must be known within $\pm 1\,^\circ\text{F}$. Scale temperature need not be known to better than $\pm 10\,^\circ\text{F}$ for comparable accuracy. Uncertainties caused by nonequilibrium temperature conditions can be avoided by installing the barometer in a uniform temperature room.

The barometer tube must be vertically aligned for accurate pressure determination. This is accomplished by a separately supported ring encircling the cistern; adjustment screws control the horizontal position.

Depression of the mercury column in commercial barometers is accounted for in the initial calibration setting at the factory. The quality of the barometer is largely determined by the bore of the glass tube. Barometers with a bore of $\frac{1}{4}$ in. are suitable for readings of about 0.01 in., whereas barometers with a bore of $\frac{1}{2}$ in. are suitable for readings down to 0.002 in.

B.6 McLeod Gage

The McLeod gage is used in making low-pressure measurements. This instrument is shown in Figure B.9 to consist of glass tubing arranged so that a sample of gas at unknown pressure can be trapped and then isothermally compressed by a rising mercury column. This amplifies the unknown pressure and allows measurement by conventional manometric means. All the mercury is initially contained in the area below the cutoff level. The gage is first exposed to the unknown gas pressure, p_1; the mercury is then raised in tube A beyond the cutoff, trapping a gas sample of initial volume $\bar{V}_1 = \bar{V} + ah_c$. The mercury is continuously forced upward until it reaches the zero level in the reference capillary B. At this time the mercury in the measuring capillary C reaches a level h, where the gas sample is at its final volume, $\bar{V}_2 = ah$, and at the amplified pressure, $p_2 = p_1 + h$. Then

$$p_1 \bar{V}_1 = p_2 \bar{V}_2 \qquad (B.20)$$

$$p_1 = \frac{ah^2}{\bar{V}_1 - ah} \qquad (B.21)$$

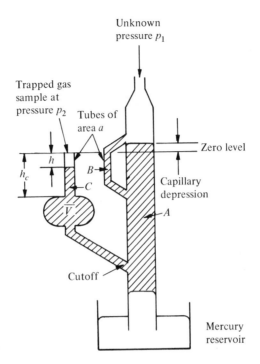

Figure B.9. McLeod gage.

If $ah \ll \bar{V}_1$, as is usually the case,

$$p_1 = \frac{h^2}{\bar{V}_1} \qquad (B.22)$$

The larger the volume ratio $(\bar{V}_1/\bar{V}_2)$, the greater will be the amplified pressure p_2 and the manometer reading h. Therefore, it is desirable that measuring tube C have a small bore. Unfortunately, for tube bores under 1 mm, the compression gain is offset by reading uncertainty caused by capillary effects.

Reference tube B is introduced to provide a meaningful zero for the measuring tube. If the zero is fixed, equation (B.22) indicates that manometer indication h varies nonlinearly with initial pressure p_1. A McLeod gage with an expanded scale at the lower pressures exhibits a higher sensitivity in this region. The McLeod pressure scale, once established, serves equally well for all the permanent gases (those whose critical pressure is appreciably below room temperature).

There are no corrections to be applied to the McLeod gage reading, but certain precautions should be taken. Moisture traps must be provided to avoid taking any condensable vapors into the gage. Such vapors occupy a larger volume at the initial low pressures than they occupy in the liquid phase at the high reading pressures. Thus the presence of condensable vapors always causes pressure readings to be too low. Capillary effects, while partially counterbalanced by using a reference capillary, can still introduce significant uncertainties since the angle of contact between mercury and glass can vary ± 30 deg. Finally, since the McLeod gage does not give continuous readings, steady-state conditions must prevail for the measurements to be useful.

The mercury piston of the McLeod gage can be actuated in a number of ways. A mechanical plunger may force the mercury up tube A, while a partial vacuum over the mercury reservoir holds the mercury below the cutoff until the gage is charged. The mercury can then be raised by bleeding dry gas into the reservoir. There are also several types of swivel gages where the mercury reservoir is located above the gage zero during charging. A 90 deg rotation of the gage causes the mercury to rise in tube A by the action of gravity. In a variation of the McLeod principle, the gas sample may be compressed between two mercury columns, thus avoiding the need for a reference capillary and a sealed-off measuring capillary.

B.7 Pressure Transducers

In the earlier sections of this appendix we have discussed five pressure standards that can be used for either calibration or the measurement of pressure in static systems. In this section we shall discuss some commonly used devices that are used for making measurements using an elastic element to convert fluid energy to mechanical energy. Such a device is known as a pressure transducer. Examples of mechanical pressure transducers having elastic elements only are dead-weight free-piston gages, manometers, Bourdon gages, bellow, and diaphragm gages.

Electrical transducers have an element that converts this displacement to an electrical signal. Active electrical transducers generate their own voltage or current output a function of displacement. Passive transducers require an external signal. The piezoelectric pickup is an example of an active electrical pressure transducer. Electric elements employed in passive electrical pressure transducers include strain gages, slide-wire potentiometers, capacitance pickups, linear differential transformers, and variable-reluctance units.

The Bourdon gage is briefly described in Chapter 2. In this transducer the elastic element is a small-volume tube that is fixed at one end but free at the other end to allow displacement under the deforming action of the pressure difference across the tube walls. In the most common model, shown in Figure B.10, a tube with an oval cross section is bent in a circular arc. Under pressure

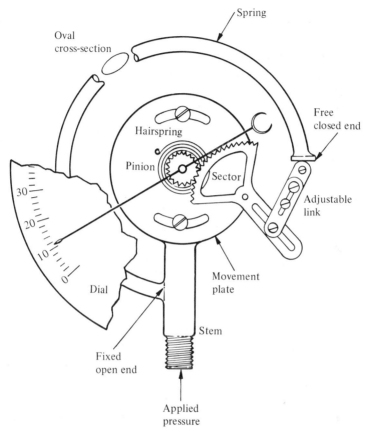

Figure B.10. *Bourdon gage.*

the tube tends to become circular, with a subsequent increase in the radius of the arc. By an almost frictionless linkage, the free end of the tube rotates a pointer over a calibrated scale to give a mechanical indication of pressure.

The reference pressure in the case containing the Bourdon tube is usually atmospheric, so that the pointer indicates gage pressures. Absolute pressures can be measured directly without evacuating the complete gage casing by biasing a sensing Bourdon tube against a reference Bourdon tube, which is evacuated and sealed as shown in Figure B.11. Bourdon gages are available for wide ranges of absolute, gage, and differential pressure measurements within a calibration uncertainty of 0.1 percent of the reading.

Another common elastic element used in pressure transducers is the bellows, shown as a gage element in Figure B.12.

In one arrangement, pressure is applied to one side of a bellows and the

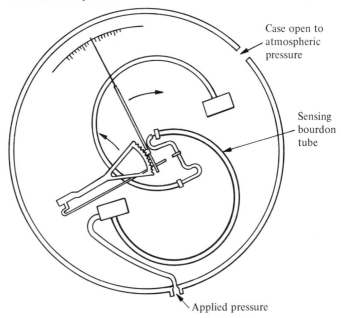

Reference bourdon tube sealed at zero absolute pressure

Case open to atmospheric pressure

Sensing bourdon tube

Applied pressure

Figure B.11. *Bourdon gage for absolute pressure measurement.*

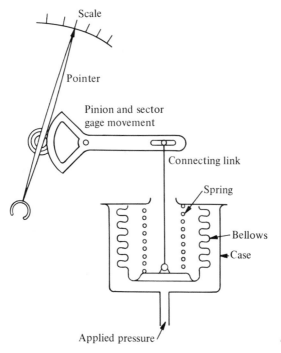

Scale

Pointer

Pinion and sector gage movement

Connecting link

Spring

Bellows

Case

Applied pressure

Figure B.12. *Bellows gage.*

338

resulting deflection is partially counterbalanced by a spring. In a differential arrangement, one pressure is applied to the inside of one sealed bellows, the other pressure is led to the inside of another sealed bellows, and the pressure difference is indicated by a pointer.

A final elastic element to be mentioned because of its widespread use in pressure transducers is the diaphragm. One such arrangement is shown in Figure B.13. Such elements may be flat, corrugated, or dished plates; the choice depends on the strength and amount of deflection desired. In high-precision instruments, a pair of diaphragms are used back to back to form an elastic capsule. One pressure is applied to the inside of the capsule; the other pressure is external. The calibration of this differential transducer is relatively independent of pressure magnitude.

Thus far we have discussed a few mechanical pressure transducers. In many applications it is more convenient to use transducer elements that depend on the change in the electrical parameters of the element as a function of the applied pressure. The only active electrical pressure transducer in common use is

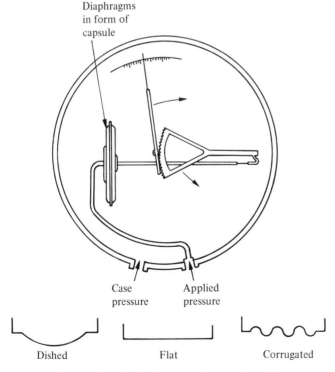

Figure B.13. Diaphragm-based pressure transducer.

the piezoelectric transducer. Sound-pressure instrumentation makes extensive use of piezoelectric pickups in such forms as hollow cylinders and disks. Piezoelectric pressure transducers are also used in measuring rapidly fluctuating or transient pressures. These transducers are difficult to calibrate by static procedures. In a recently introduced technique called electrocalibration the transducer is calibrated by electric field excitation rather than by physical pressure. The most common passive electrical pressure transducers are the variable resistance types.

The strain gage is probably the most used pressure transducer element. Strain gages operate on the principle that the electrical resistance of a wire varies with its length under load. In unbonded strain gages, four wires run between electrically insulated pins located on a fixed frame and other pins located on a movable armature, as shown in Figure B.14. The wires are installed

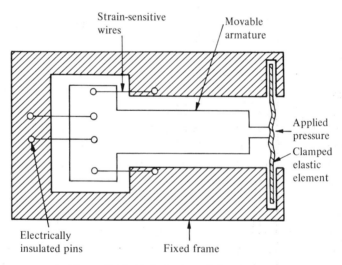

Figure B.14. *Typical unbounded strain gage.*

under tension and form the legs of a bridge circuit. Under pressure the elastic element (usually a diaphragm) displaces the armature, causing two of the wires to elongate while reducing the tension in the remaining two wires. The resistance change causes a bridge imbalance proportional to the applied pressure.

The bonded strain gage takes the form of a fine wire filament, set in cloth, paper, or plastic and fastened by a suitable cement to a flexible plate, which takes the load of the elastic element. This is shown in Figure B.15. Two similar strain gage elements are often connected in a bridge circuit to balance unavoidable temperature effects. The nominal bridge output impedance of most strain

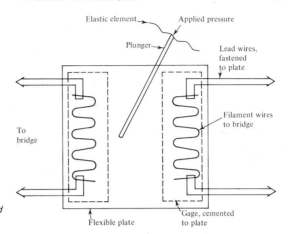

Elastic element⟶   Applied pressure

Plunger⟶

Lead wires,
fastened
to plate

Filament wires
to bridge

To
bridge

Figure B.15. *Typical bonded
strain gage.*

Flexible plate   Gage, cemented
to plate

gage pressure transducers is 350 Ω, nominal excitation voltage is 10 V (ac or dc), and natural frequency can be as high as 50 Hz. Transducer resolution is infinite, and the usual calibration uncertainty of such gages is within 1 percent of full scale.

Many other forms of electrical pressure transducers are in use in industry, and only a few of these will be discussed briefly. These elements fall under the categories of potentiometer, variable capacitance, linear variable differential transformer (LVDT), and variable reluctance transducers. The potentiometer types are those that operate as variable resistance pressure transducers. In one arrangement, the elastic element is a helical Bourdon tube, while a precision wire-wound potentiometer serves as the electric element. As pressure is applied to the open end of the Bourdon tube it unwinds, causing the wiper (connected directly to the closed end of the tube) to move over the potentiometer.

In the variable capacitance pressure transducer, the elastic element is usually a metal diaphragm that serves as one plate of a capacitor. Under an applied pressure the diaphragm moves with respect to a fixed plate. By means of a suitable bridge circuit, the variation in capacitance can be measured and related to pressure by calibration.

The electric element in an LVDT is made up of three coils mounted in a common frame to form the device shown in Figure B.16. A magnetic core centered in the coils is free to be displaced by a bellows, Bourdon, or diaphragm elastic element. The center coil, the primary winding of the transformer, has an ac excitation voltage impressed across it. The two outside coils form the secondaries of the transformer. When the core is centered, the induced voltages in the two outer coils are equal and 180 deg out of phase; this represents the zero pressure position. However, when the core is displaced by the action of an applied pressure, the voltage induced in one secondary increases, while that in

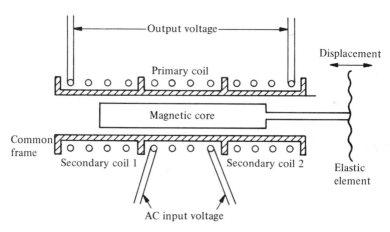

Figure B.16. *Linear variable differential transformer.*

the other decreases. The output-voltage difference varies essentially linearly with pressure for the small core displacements allowed in these transducers. The voltage difference is measured and related to the applied pressure by calibration.

The last electric element that will be discussed is the variable reluctance pressure transducer. The basic element of the variable reluctance pressure transducer is a movable magnetic vane in a magnetic field. In one type the elastic element is a flat magnetic diaphragm located between two magnetic output coils. Displacement of the diaphragm changes the inductance ratio between the output coils and results in an output voltage proportional to pressure.

In this appendix the pressure elements that have been discussed were primarily for the measurement of pressures in systems that are at rest and in which the fluid is in equibrium. The measurement of pressure in systems in which the fluid is moving requires the further definition of pressure as well as an understanding of the pressure gradients, turbulence, viscous effects, and time-dependent effects that occur in such systems. The usual elements for making these measurements are the static or wall tap, static tubes, aerodynamic probes, Pitot tubes, Pitot-static tubes, etc. The principles of these devices have been discussed elsewhere in this text. The student desiring further information is referred to the article by R. P. Benedict cited earlier in this appendix or the book by Benedict (*Fundamentals of Temperature, Pressure and Flow Measurements* by R. P. Benedict, John Wiley & Sons, Inc., New York. 1969) for an extensive bibliography on this subject.

SELECTED PHYSICAL DATA

Table C.1 Conversion Tables†

English to Metric

To Change	To	Multiply by
Inches	Millimeters	25.4
Inches	Centimeters	2.54
Inches	Meters	0.0254
Feet	Meters	0.3048
Miles	Kilometers	1.609347
Square inches	Square centimeters	6.452
Square feet	Square meters	0.0929
Cubic inches	Cubic centimeters	16.3872
Cubic feet	Cubic meters	0.02832
U. S. gallons	Liters	3.7854
Pounds	Kilograms	0.45359
Ounces avoir	Grams	28.3495
B.T.U.	Calories	252
B.T.U. per hour per square feet per degree F	Kilogram calories per hour per square meter per degree C	4.88
Pounds per square inch	Kilogram per square centimeter	0.0703

Metric to English

To Change	To	Multiply by
Millimeters	Inches	0.03937
Centimeters	Inches	0.39371
Meters	Inches	39.371
Meters	Feet	3.281
Kilometers	Miles	0.62137
Square centimeters	Square inches	0.1550
Square meters	Square feet	10.7649
Cubic centimeters	Cubic inches	0.061
Cubic meters	Cubic feet	35.314
Liters	U.S. gallons	0.26417
Kilograms	Pounds	2.20462
Grams	Ounces avoir	0.03527
Calories	B.T.U	0.003968
Kilogram calories per hour per square meter per degree C	B.T.U. per hour per square foot per degree F	0.205
Kilograms per square centimeter	Pounds per square inch	14.223

Measure

To Change	To	Multiply by
Cubic feet	Cubic inches	1728
Cubic inches	Cubic feet	0.00058
Cubic feet	Gallons	7.480
Gallons	Cubic feet	0.1337
Cubic inches	Gallons	0.00433
Gallons	Cubic inches	231
Barrels	Gallons	42
Gallons	Barrels	0.0238
Imperial gallons	U.S. gallons	1.2009
U.S. gallons	Imperial gallons	0.8326
Feet	Inches	12
Inches	Feet	0.0833
Square feet	Square inches	144
Square inches	Square feet	0.00695
Long tons	Pounds	2240
Short tons	Pounds	2000
Long tons	Short tons	1.12

Weight

Volume	Weight (Pounds)
1 cubic foot of water	62.4 (@32°F)
1cubic inch of water	0.0361(@32°F)
1gallon of water	8.33(@32°F)
1cubic foot of air	0.0763(@60°F-29.92 in. Hg)
1cubic inch of steel	0.284
1cubic foot of brick (bldg)	112-120
1cubic foot of brick (fire)	145-150
1cubic foot of coal	80-100
1cubic foot of coke	24- 30
1cubic foot of concrete	120-140
1cubic foot of earth	70-120
1cubic foot of gravel	90-120
1cubic foot of wood	30- 60

Pressure (at 32°F)

To Change	To	Multiply by
Inches of water	Pounds per square inch	0.03612
Pounds per square inch	Inches of water	27.686
Feet of water	Pounds per square inch	0.4334
Pounds per square inch	Feet of water	2.307
Inches of mercury	Pounds per square inch	0.4912
Pounds per square inch	Inches of mercury	2.036
Atmospheres	Pounds per square inch	14.696
Pounds per square inch	Atmospheres	0.06804

Density

To Change	To	Multiply by
Pounds per cubic foot	Kilograms per cubic meter	16.0184
Kilograms per cubic meter	Pounds per cubic foot	0.06243

Power

To Change	To	Multiply by
Horsepower	Kilowatts	0.746
Kilowatts	Horsepower	1.3404
B.T.U	Foot-pounds	778.3
Foot-pounds	B.T.U	0.001285
B.T.U	Horsepower hours	0.0003927
Horsepower hours	B.T.U	2544.1
B.T.U	Kilowatts hours	0.0002928
Kilowatt hours	B.T.U	3412.75

†Data taken with permission from *Catalog #57*, the Walworth Co., New York, 1957.

Table C.2 Properties of pipe†

Schedules, Wall Thicknesses, and Weights

Conforming to ASA Standard B36.10, 1950, for Wrought Steel and Wrought Iron Pipe

Nominal Pipe Size (in.)	Outside Diameter, D(in.)	Wall Thickness, t(in.)	Inside Diameter d(in.)	Inside Diameter d²(squared)	Inside Diameter d⁵(fifth power)	Area of Metal (in.³)	Internal Cross-Sectional Area in.²	Internal Cross-Sectional Area ft²	External Surface (f²)	Moment of Inertia (in.⁴)	Weight (Pounds) of Pipe (per ft)	Weight (Pounds) of Water (per ft of pipe)
					Schedule 10							
14 o.d.	14.0	0.250	13.50	182.25	448,403	10.80	143.14	0.994	3.665	255.3	36.71	62.03
16 o.d.	16.0	0.250	15.50	240.25	894,660	12.37	188.69	1.310	4.189	383.7	42.05	81.74
18 o.d.	18.0	0.250	17.50	306.25	1,641,309	13.94	240.53	1.670	4.712	549.1	47.39	104.21
20 o.d.	20.0	0.250	19.50	380.25	2,819,505	15.51	298.65	2.074	5.236	756.4	52.73	129.42
24 o.d.	24.0	0.250	23.50	552.25	7,167,030	18.65	433.74	3.012	6.283	1,315.0	63.41	187.95
30 o.d.	30.0	0.312	29.376	862.95	21,875,768	29.10	677.76	4.707	7.854	3,206.0	98.93	293.72
					Schedule 20							
8	8.625	0.250	8.125	66.02	35,409	6.57	51.85	0.3601	2.258	57.72	22.36	22.47
10	10.75	0.250	10.25	105.06	113,141	8.24	82.52	0.5731	2.814	113.7	28.04	35.76
12	12.75	0.250	12.25	150.06	275,855	9.82	117.86	0.8185	3.338	191.8	33.38	51.07
14 o.d.	14.0	0.312	13.376	178.92	428,185	13.42	140.52	0.975	3.665	314.4	45.68	60.89
16 o.d.	16.0	0.312	15.376	236.42	859,442	15.38	185.69	1.290	4.189	473.2	52.36	80.50
18 o.d.	18.0	0.312	17.376	301.92	1,583,978	17.34	237.13	1.647	4.712	678.2	59.03	102.77
20 o.d.	20.0	0.375	19.25	370.56	2,643,344	23.12	291.04	2.021	5.236	1113	78.60	125.67
24 o.d.	24.0	0.375	23.25	540.56	6,793,832	27.83	424.56	2.948	6.283	1942	94.62	183.95
30 o.d.	30.0	0.500	29.00	841.0	20,511,149	46.34	660.52	4.587	7.854	5,042	157.53	286.23
					Schedule 30							
8	8.625	0.277	8.071	65.14	34,248	7.26	51.16	0.3553	2.258	63.35	24.70	22.17
10	10.75	0.307	10.136	102.74	106,987	10.07	80.69	0.5603	2.814	137.4	34.24	34.96

12	12.75	0.330	12.09	146.17	258,304	12.87	114.80	0.7972	3.338	248.4	43.77	49.74
14 o.d.	14.0	0.375	13.25	175.56	408,394	16.05	137.88	0.9575	3.665	372.8	54.57	59.75
16 o.d.	16.0	0.375	15.25	232.56	824,801	18.41	182.65	1.268	4.189	562.1	62.58	79.12
18 o.d.	18.0	0.437	17.126	293.30	1,473,261	24.11	230.36	1.600	4.712	930.3	82.06	99.84
20 o.d.	20.0	0.500	19.0	361.00	2,476,099	30.63	283.53	1.969	5.236	1,457	104.13	122.87
24 o.d.	24.0	0.562	22.876	523.31	6,264,703	41.39	411.00	2.854	6.283	2,843	140.80	178.09
30 o.d.	30.0	0.625	28.75	826.56	19,642,160	57.68	649.18	4.508	7.854	6,224	196.08	281.30

Schedule 40

$\frac{1}{8}$	0.405	0.068§	0.269	0.0724	0.00141	0.072	0.0569	0.00040	0.106	0.001064	0.24	0.0250
$\frac{1}{4}$	0.540	0.088§	0.364	0.1325	0.00639	0.125	0.1041	0.00072	0.141	0.003312	0.42	0.0449
$\frac{3}{8}$	0.675	0.091§	0.493	0.2430	0.02912	0.167	0.1909	0.00133	0.177	0.007291	0.57	0.0830
$\frac{1}{2}$	0.840	0.109§	0.622	0.3869	0.09310	0.250	0.3039	0.00211	0.220	0.01709	0.85	0.1317
$\frac{3}{4}$	1.050	0.113§	0.824	0.679	0.3799	0.333	0.5333	0.00371	0.275	0.03704	1.13	0.2315
1	1.315	0.133§	1.049	1.100	1.270	0.494	0.8639	0.00600	0.344	0.08734	1.68	0.3744
$1\frac{1}{4}$	1.660	0.140§	1.380	1.904	5.005	0.669	1.495	0.01040	0.435	0.1947	2.27	0.6490
$1\frac{1}{2}$	1.900	0.145§	1.610	2.592	10.82	0.799	2.036	0.01414	0.497	0.3099	2.72	0.8823
2	2.375	0.154§	2.067	4.272	37.72	1.075	3.356	0.02330	0.622	0.666	3.65	1.454
$2\frac{1}{2}$	2.875	0.203§	2.469	6.096	91.75	1.704	4.788	0.03322	0.753	1.530	5.79	2.073
3	3.5	0.216§	3.068	9.413	271.8	2.228	7.393	0.05130	0.916	3.017	7.58	3.201
$3\frac{1}{2}$	4.0	0.226§	3.548	12.59	562.2	2.680	9.888	0.06870	1.047	4.788	9.11	4.287
4	4.5	0.237§	4.026	16.21	1,058	3.173	12.73	0.08840	1.178	7.233	10.79	5.516
5	5.563	0.258§	5.047	25.47	3,275	4.304	20.01	0.1390	1.456	15.16	14.62	8.674
6	6.625	0.280§	6.065	36.78	8,206	5.584	28.89	0.2006	1.734	28.14	18.97	12.52
8	8.625	0.322§	7.981	63.70	32,380	8.396	50.03	0.3474	2.258	72.49	25.55	21.68
10	10.75	0.365§	10.02	100.4	101,000	11.90	78.85	0.5475	2.814	160.7	40.48	34.16
12	12.75	0.406	11.938	142.5	242,470	15.77	111.93	0.7773	3.338	300.3	53.53	48.50
14 o.d.	14.0	0.437	13.126	172.3	389,638	18.61	135.32	0.9397	3.665	429.1	63.37	58.64
16 o.d.	16.0	0.500	15.000	225.0	759,375	24.35	176.72	1.2272	4.189	731.9	82.77	76.58
18 o.d.	18.0	0.562	16.876	284.8	1,368,820	30.79	223.68	1.5533	4.712	1172	104.75	96.93

Nominal Pipe Size (in.)	Outside Diameter D(in.)	Wall Thickness t(in.)	Inside Diameter d(in.)	Inside Diameter d²(squared)	Inside Diameter d⁵(fifth power)	Area of Metal (in.³)	Internal Cross-Sectional Area in.²	Internal Cross-Sectional Area ft²	External Surface (ft²)	Moment of Inertia (in.⁴)	Weight (Pounds) of Pipe (per ft)	Weight (Pounds) of Water (per ft of pipe)
20 o.d.	20.0	0.593	18.814	354.0	2,357,244	36.15	278.00	1.9305	5.236	1703	122.91	120.46
24 o.d.	24.0	0.687	22.626	511.9	5,929,784	50.31	402.07	2.7921	6.283	3424	171.17	174.23
Schedule 60												
8	8.625	0.406	7.813	61.04	29,113	10.48	47.94	0.3329	2.258	88.73	35.64	20.77
10	10.75	0.500¶	9.75	95.06	88,110	16.10	74.66	0.5185	2.814	212.0	54.74	32.35
12	12.75	0.562	11.626	135.16	212,399	21.52	106.16	0.7372	3.338	400.4	73.16	46.00
14 o.d.	14.0	0.593	12.814	164.20	345,480	24.98	128.96	0.8956	3.665	562.3	84.91	55.86
16 o.d.	16.0	0.656	14.688	215.74	683,618	31.62	169.44	1.1766	4.189	932.4	107.50	73.42
18 o.d.	18.0	0.750	16.500	272.25	1,222,981	40.64	213.83	1.4849	4.712	1,515	138.17	92.80
20 o.d.	20.0	0.812	18.376	337.68	2,095,342	48.95	265.21	1.8417	5.236	2,257	166.40	114.92
24 o.d.	24.0	0.968	22.064	486.82	5,229,029	70.04	382.35	2.6552	6.283	4,654	238.11	165.94
Schedule 80												
1/8	0.405	0.095¶	0.215	0.0462	0.000459	0.093	0.0363	0.00025	0.106	0.001216	0.31	0.0157
1/4	0.540	0.119¶	0.302	0.0912	0.002513	0.157	0.0716	0.00050	0.141	0.003766	0.54	0.031
3/8	0.675	0.126¶	0.423	0.1789	0.01354	0.217	0.1405	0.00098	0.177	0.008619	0.74	0.0609
1/2	0.840	0.147¶	0.546	0.2981	0.04852	0.320	0.2341	0.00163	0.220	0.02008	1.09	0.1013
3/4	1.050	0.154¶	0.742	0.5506	0.2249	0.433	0.4324	0.00300	0.275	0.04479	1.47	0.1875
1	1.315	0.179¶	0.957	0.9158	0.8027	0.639	0.7193	0.00499	0.344	0.1056	2.17	0.3112
1¼	1.660	0.191¶	1.278	1.633	3.409	0.881	1.283	0.00891	0.435	0.2418	3.00	0.5553
1½	1.900	0.200¶	1.500	2.250	7.594	1.068	1.767	0.01225	0.498	0.3912	3.63	0.7648
2	2.375	0.218¶	1.939	3.760	27.41	1.477	2.953	0.02050	0.622	0.8679	5.02	1.279
2½	2.875	0.276¶	2.323	5.396	67.64	2.254	4.238	0.02942	0.753	1.924	7.66	1.834
3	3.5	0.300¶	2.900	8.410	205.1	3.016	6.605	0.04587	0.917	3.894	10.25	2.859
3½	4.0	0.318¶	3.364	11.32	430.8	3.678	8.891	0.06170	1.047	6.280	12.51	3.847

4	0.337	3.826	14.64	819.8	4.407	11.50	0.07986	1.178	9.610	14.98	4.976
5	0.375	4.813	23.16	2,583	6.112	18.19	0.1263	1.456	20.67	20.78	7.875
6	0.432	5.761	33.19	6,346	8.405	26.07	0.1810	1.734	40.49	28.57	11.29
8	0.500	7.625	58.14	25,775	12.76	45.66	0.3171	2.257	105.7	43.39	19.79
10	0.593	9.564	91.47	80,020	18.92	71.84	0.4989	2.817	244.8	64.33	31.13
12	0.687	11.376	129.41	190,523	26.03	101.64	0.7958	3.338	475.1	88.51	44.04
14 o.d.	0.750	12.500	156.25	305,176	31.22	122.72	0.8522	3.665	687.3	106.13	53.18
16 o.d.	0.843	14.314	204.89	600,904	40.14	160.92	1.1175	4.189	1,156	136.46	69.73
18 o.d.	0.937	16.125	260.05	1,090,518	50.23	204.24	1.4183	4.712	1,833	170.75	88.50
20 o.d.	1.031	17.938	321.77	1,857,248	61.44	252.72	1.7550	5.236	2,772	208.87	109.51
24 o.d.	1.218	21.564	465.01	4,662,798	87.17	365.22	2.5362	6.283	5,672	296.36	158.26
Schedule 100											
8	0.593	7.439	55.34	22,781	14.96	43.46	0.3018	2.258	121.3	50.87	18.83
10	0.718	9.314	86.75	69,357	22.63	68.13	0.4732	2.814	286.1	76.93	29.53
12	0.843	11.064	122.41	165,791	31.53	96.14	0.6677	3.338	561.6	107.20	41.66
14 o.d.	0.937	12.126	147.04	262,173	38.45	115.49	0.8020	3.665	824.4	130.73	50.04
16 o.d.	1.031	13.938	194.27	526,020	48.48	152.58	1.0596	4.189	1,364	164.83	66.12
18 o.d.	1.156	15.688	246.11	950,250	61.17	193.30	1.3423	4.712	2,180	207.96	83.76
20 o.d.	1.281	17.438	304.08	1,612,398	75.34	238.82	1.6585	5.236	3,316	256.10	103.65
24 o.d.	1.531	20.938	438.40	4,024,179	108.07	344.32	2.3911	6.283	6,853	367.40	149.43
Schedule 120											
4	0.438	3.625	13.15	626.8	5.578	10.33	0.0717	1.178	11.65	19.01	4.47
5	0.500	4.563	20.82	1,978	7.953	16.35	0.1136	1.456	25.73	27.04	7.09
6	0.562	5.501	30.26	5,037	10.705	23.77	0.1650	1.734	49.61	36.39	10.30
8	0.718	7.189	51.68	19,202	17.84	40.59	0.2819	2.257	140.5	60.63	17.59
10	0.843	9.064	82.16	61,179	26.24	64.53	0.4481	2.817	324.2	89.20	27.96
12	1.000	10.750	115.56	143,563	36.91	90.76	0.6303	3.338	641.6	125.49	39.33
14 o.d.	1.093	11.814	139.57	230,134	44.32	109.62	0.7612	3.665	929.8	150.67	47.57
16 o.d.	1.218	13.564	183.98	459,133	56.56	144.50	1.0035	4.189	1,555	192.29	62.62

Nominal Pipe Size (in.)	Outside Diameter D(in.)	Wall Thickness t(in.)	Inside Diameter d(in.)	Inside Diameter d²(squared)	Inside Diameter d⁵(fifth power)	Area of Metal (in.³)	Internal Cross-Sectional Area in.²	Internal Cross-Sectional Area ft²	External Surface (ft²)	Moment of Inertia (in.⁴)	Weight (Pounds) of Pipe (per ft)	Weight (Pounds) of Water (per ft of pipe)
18 o.d.	18.0	1.375	15.250	232.56	824,783	71.82	182.65	1.2684	4.712	2,499	244.14	79.27
20 o.d.	20.0	1.500	17.000	289.00	1,419,857	87.18	226.98	1.5762	5.236	3,754	296.37	98.35
24 o.d.	24.0	1.812	20.376	415.18	3,512,301	126.31	326.08	2.2644	6.283	7,827	429.39	141.52
Schedule 140												
8	8.625	0.812	7.001	49.01	16,819	19.93	38.50	0.2673	2.257	153.7	67.76	16.68
10	10.75	1.000	8.750	76.56	51,291	30.63	60.13	0.4176	2.817	367.8	104.13	26.06
12	12.75	1.125	10.500	110.25	127,628	41.08	86.59	0.6013	3.338	700.5	139.68	37.52
14 o.d.	14.0	1.250	11.500	132.25	201,136	50.07	103.87	0.7213	3.665	1,027	170.22	45.01
16 o.d.	16.0	1.438	13.125	172.29	389,670	65.74	135.32	0.9397	4.189	1,760	223.50	58.64
18 o.d.	18.0	1.562	14.876	221.30	728,502	80.66	173.80	1.2070	4.712	2,749	274.23	75.32
20 o.d.	20.0	1.750	16.500	272.25	1,222,981	100.33	213.82	1.4849	5.236	4,216	341.10	92.66
24 o.d.	24.0	2.062	19.876	395.09	3,102,022	142.11	310.28	2.1547	6.283	8,625	483.13	134.45
Schedule 160												
½	0.840	0.187	0.466	0.2172	0.002197	0.3836	0.1706	0.00118	0.220	0.02212	1.30	0.074
¾	1.050	0.218	0.614	0.3770	0.08726	0.5698	0.2961	0.00206	0.275	0.05269	1.94	0.130
1	1.315	0.250	0.815	0.6642	0.3596	0.8365	0.5217	0.00362	0.344	0.1251	2.84	0.230
1¼	1.660	0.250	1.160	1.346	2.100	1.107	1.057	0.00734	0.435	0.2839	3.76	0.46
1½	1.900	0.281	1.338	1.790	4.288	1.429	1.406	0.00976	0.498	0.4824	4.86	0.61
2	2.375	0.343	1.689	2.853	13.74	2.190	2.241	0.01556	0.622	1.162	7.44	0.97
2½	2.875	0.375	2.125	4.516	43.33	2.945	3.546	0.02463	0.753	2.353	10.01	1.54
3	3.5	0.438	2.625	6.896	124.9	4.205	5.416	0.03761	0.917	5.032	14.32	2.35
3½	4.0	—	—	—	—	—	—	—	—	—	—	—
4	4.5	0.531	3.438	11.82	480.3	6.621	9.283	0.06447	1.178	13.27	22.51	4.02
5	5.563	0.625	4.313	18.60	1,492	9.696	14.61	0.1015	1.456	30.03	32.96	6.33

	6.625	0.718	5.189	26.93	3,762	13.32	21.15	0.1469	1.734	58.97	45.30	9.16
6	6.625	0.718	5.189	26.93	3,762	13.32	21.15	0.1469	1.734	58.97	45.30	9.16
8	8.625	0.906	6.813	46.42	14,679	21.97	36.46	0.2532	2.257	165.9	74.69	15.80
10	10.75	1.125	8.500	72.25	44,371	34.02	56.75	0.3941	2.817	399.3	115.65	24.59
12	12.75	1.312	10.126	102.54	106,461	47.14	80.53	0.5592	3.338	781.1	160.27	34.89
14 o.d.	14.0	1.406	11.188	125.17	175,292	55.63	98.31	0.6827	3.665	1,117	189.12	42.60
16 o.d.	16.0	1.593	12.814	164.20	345,486	72.10	128.96	0.8955	4.189	1,894	245.11	55.97
18 o.d.	18.0	1.781	14.438	208.46	627,412	90.75	163.72	1.1369	4.712	3,021	308.51	71.05
20 o.d.	20.0	1.968	16.064	258.05	1,069,699	111.49	202.67	1.4074	5.236	4,586	379.01	87.96
24 o.d.	24.0	2.343	19.314	373.03	2,687,570	159.41	292.98	2.0345	6.283	9,458	541.94	127.15

† Data taken with permission from *Catalog #57*, the Walworth Co., New York, 1957.
‡ This column also represents the contents in cubic feet per foot of length.
§ These thicknesses are identical with those listed in ASA B36.10—1950 for standard wall pipe.
¶ These thicknesses are identical with those listed in ASA B36.10—1950 for extra strong wall pipe.

Table C.3 Average Properties of Tubes[†]

Diameter		Thickness		External			Internal				
External (in.)	Internal (in.)	BWG Gage	NOM Wall (in.)	Circumference (in.)	Surface per Lineal Foot (ft²)	Lineal Feet of Tube per Square Foot of Surface	Transverse Area (in.²)	Volume or Capacity per Lineal Foot In.³	Ft.³	U.S. gal	Length of Tube Containing One Cubic Foot
$\frac{5}{8}$	0.527	18	0.049	1.9635	0.1636	6.1115	0.218	2.616	0.0015	0.011	661
	0.495	16	0.065	1.9635	0.1636	6.1115	0.193	2.316	0.0013	0.010	746
	0.459	14	0.083	1.9635	0.1636	6.1115	0.166	1.992	0.0011	0.009	867
$\frac{3}{4}$	0.652	18	0.049	2.3562	0.1963	5.0930	0.334	4.008	0.0023	0.017	431
	0.620	16	0.065	2.3562	0.1963	0.0930	0.302	3.624	0.0021	0.016	477
	0.584	14	0.083	2.3562	0.1963	0.0930	0.268	3.216	0.0019	0.014	537
	0.560	13	0.095	2.3562	0.1963	5.0930	0.246	2.952	0.0017	0.013	585
1	0.902	18	0.049	3.1416	0.2618	3.8197	0.639	7.668	0.0044	0.033	225
	0.870	16	0.065	3.1416	0.2618	3.8197	0.595	7.140	0.0041	0.031	242
	0.834	14	0.083	3.1416	0.2618	3.8197	0.546	6.552	0.0038	0.028	264
	0.810	13	0.095	3.1416	0.2618	3.8197	0.515	6.180	0.0036	0.027	280
$1\frac{1}{4}$	1.152	18	0.049	3.9270	0.3272	3.0558	1.075	12.90	0.0075	0.056	134
	1.120	16	0.065	3.9270	0.3272	3.0558	0.985	11.82	0.0068	0.051	146
	1.084	14	0.083	3.9270	0.3272	3.0558	0.923	11.08	0.0064	0.048	156
	1.060	13	0.095	3.9270	0.3272	3.0558	0.882	10.58	0.0061	0.046	163
	1.032	12	0.109	3.9270	0.3272	3.0558	0.836	10.03	0.0058	0.043	172
$1\frac{1}{2}$	1.402	18	0.049	4.7124	0.3927	2.5465	1.544	18.53	0.0107	0.080	93
	1.370	16	0.065	4.7124	0.3927	2.5465	1.474	17.69	0.0102	0.076	98
	1.334	14	0.083	4.7124	0.3927	2.5465	1.398	16.78	0.0097	0.073	103
	1.310	13	0.095	4.7124	0.3927	2.5465	1.343	16.12	0.0093	0.070	107
	1.282	12	0.109	4.7124	0.3927	2.5465	1.292	15.50	0.0090	0.067	111
$1\frac{3}{4}$	1.620	16	0.065	5.4978	0.4581	2.1827	2.061	24.73	0.0143	0.107	70
	1.584	14	0.083	5.4978	0.4581	2.1827	1.971	23.65	0.0137	0.102	73
	1.560	13	0.095	5.4978	0.4581	2.1827	1.911	22.94	0.0133	0.099	75
	1.532	12	0.109	5.4978	0.4581	2.1827	1.843	22.12	0.0128	0.096	78
	1.490	11	0.120	5.4978	0.4581	2.1827	1.744	20.92	0.0121	0.090	83
2	1.870	16	0.065	6.2832	0.5236	1.9099	2.746	32.96	0.0191	0.143	52
	1.834	14	0.083	6.2832	0.5236	1.9099	2.642	31.70	0.0183	0.137	55
	1.810	13	0.095	6.2832	0.5236	1.9099	2.573	30.88	0.0179	0.134	56
	1.782	12	0.109	6.2832	0.5236	1.9099	2.489	29.87	0.0173	0.129	58
	1.760	11	0.120	6.2832	0.5236	1.9099	2.433	29.20	0.0169	0.126	59

[†]Reproduced with permission from *Principles of Heat Transfer* by Frank Kreith, International Textbook Company, Scranton, Pa., 1958, p. 541.

Table C.4 Expansion Coefficient or Liquids at 1 atm Pressure and 70°F†

Liquid	°F⁻¹
Water	1.2×10^{-4}
Ethyl alcohol	6.21×10^{-4}
Freon 12	1.40×10^{-4}
Mercury	1.01×10^{-4}
Silicone oil	4.80×10^{-4}
Petroleum oil	4.0×10^{-4}

†Reproduced with permission from *Thermodynamics of Fluid Flow* by N. A. Hall, Prentice-Hall, Inc., Englewood Cliffs, N. J. 1951, p. 11.

Table C.5 Gas-Constant Values†

Substance	Symbol	M	R (ft lb/ lb$_m$ °R)	C$_p$ (Btu/lb$_m$ °R) at 77°F	C$_v$ (Btu/lb$_m$ °R) at 77°F	K (C$_p$/C$_v$)
Acetylene	C$_2$H$_2$	26.038	59.39	0.4030	0.3267	1.234
Air	—	28.967	53.36	0.2404	0.1718	1.399
Ammonia	NH$_3$	17.032	90.77	0.5006	0.3840	1.304
Argon	A	39.944	38.73	0.1244	0.0746	1.668
Benzene	C$_6$H$_6$	78.114	19.78	0.2497	0.2243	1.113
n-Butane	C$_4$H$_{10}$	58.124	26.61	0.4004	0.3662	1.093
Isobutane	C$_4$H$_{10}$	58.124	26.59	0.3979	0.3637	1.094
1-Butene	C$_4$H$_8$	56.108	27.545	0.3646	0.3282	1.111
Carbon dioxide	CO$_2$	44.011	35.12	0.2015	0.1564	1.288
Carbon monoxide	CO	28.011	55.19	0.2485	0.1776	1.399
Carbon tetrachloride	CCl$_4$	153.839				
n-Deuterium	D$_2$	4.029				
Dodecane	C$_{12}$H$_{26}$	170.340	9.074	0.3931	0.3814	1.031
Ethane	C$_2$H$_6$	30.070	51.43	0.4183	0.3522	1.188
Ethyl ether	C$_4$H$_{10}$O	74.124				
Ethylene	C$_2$H$_4$	28.054	55.13	0.3708	0.3000	1.236
Freon, F-12	CCl$_2$F$_2$	120.925	12.78	0.1369	0.1204	1.136
Helium	He	4.003	386.33	1.241	0.7446	1.667
n-Heptane	C$_7$H$_{16}$	100.205	15.42	0.3956	0.3758	1.053
n-Hexane	C$_6$H$_{14}$	86.178	17.93	0.3966	0.3736	1.062
Hydrogen	H$_2$	2.016	766.53	3.416	2.431	1.405
Hydrogen sulfide	H$_2$S	34.082				
Mercury	Hg	200.610				
Methane	CH$_4$	16.043	96.40	0.5318	0.4079	1.304
Methyl fluoride	CH$_3$F	34.035				
Neon	Ne	20.183	76.58	0.2460	0.1476	1.667
Nitric oxide	NO	30.008	51.49	0.2377	0.1715	1.386
Nitrogen	N$_2$	28.016	55.15	0.2483	0.1774	1.400
Octane	C$_8$H$_{18}$	114.232	13.54	0.3949	0.3775	1.046
Oxygen	O$_2$	32.000	48.29	0.2191	0.1570	1.396
n-Pentane	C$_5$H$_{12}$	72.151	21.42	0.3980	0.3705	1.074
Isopentane	C$_5$H$_{12}$	72.151	21.42	0.3972	0.3697	1.074
Propane	C$_3$H$_8$	44.097	35.07	0.3982	0.3531	1.128
Propylene	C$_3$H$_6$	42.081	36.72	0.3627	0.3055	1.187
Sulfur dioxide	SO$_2$	64.066	24.12	0.1483	0.1173	1.264
Water vapor	H$_2$O	18.016	85.80	0.4452	0.3349	1.329
Xenon	Xe	131.300	11.78	0.03781	0.02269	1.667

†Reproduced with permission from *Concepts of Thermodynamics* by E. F. Obert, McGraw-Hill Book Company, Inc., New York, 1960, p. 502.

Table C.6 Absolute and Kinematic Viscosities of Water†

°F	$\mu \times 10^5$ (lb$_f$ sec/ft^2)	$v \times 10^5$ (ft^2/sec)	°F	$\mu \times 10^5$ (lb$_f$ sec/ft^2)	$v \times 10^5$ (ft^2/sec)
32	3.75	1.93	120	1.17	0.609
35	3.54	1.82	125	1.12	0.582
40	3.23	1.66	130	1.08	0.562
45	2.97	1.53	135	1.02	0.534
50	2.73	1.41	140	0.981	0.514
55	2.53	1.30	145	0.940	0.493
60	2.35	1.22	150	0.899	0.472
65	2.24	1.13	155	0.868	0.457
70	2.04	1.05	160	0.837	0.440
75	1.92	0.988	165	0.806	0.426
80	1.80	0.929	170	0.776	0.411
85	1.68	0.870	175	0.750	0.397
90	1.60	0.825	180	0.725	0.384
95	1.51	0.782	185	0.701	0.372
100	1.43	0.738	190	0.679	0.362
105	1.35	0.698	195	0.657	0.351
110	1.29	0.668	200	0.637	0.341
115	1.23	0.637	212	0.593	0.318

†Reproduced with permission from *Fluid Mechanics* by Arthur G. Hansen, John Wiley & Sons, Inc., New York, 1967, p. 486.

Table C.7 Surface Tension of Water Exposed to Air or Its Own Vapor†

°F	Surface Tension σ (lb$_f$/ft)
32	5.2×10^{-3}
40	5.1
60	5.0
80	4.9
100	4.8
120	4.7
140	4.5
160	4.4
180	4.3
200	4.1
212	4.0

†Reproduced with permission from *Fluid Mechanics* by Arthur G. Hansen, John Wiley & Sons, Inc., New York, 1967, p. 486.

Table C.8 Surface Tension of Various Liquids in Contact with Air[†]

Substance (in contact with air)	°F	Surface Tension σ (lb_f/ft)
Mercury	68	0.0324
Benzene	68	0.00198
Carbon tetrachloride	68	0.00184
Glycerine	68	0.00482
Ethyl alcohol	68	0.00153
Methyl alcohol	68	0.00155

[†]Reproduced with permission from *Fluid Mechanics* by Arthur G. Hansen, John Wiley & Sons, Inc., New York, 1967, p. 487.

Table C.9 Properties of the Standard Atmosphere[†]

Altitude (ft)	Temperature (°F)	Absolute Pressure (lb_f/ft^2)	γ (lb/ft^3)	Speed of Sound (ft/sec)
0	59	2116	0.0765	1117
5,000	41	1761	0.0660	1098
10,000	23	1455	0.0566	1078
15,000	6	1194	0.0482	1058
20,000	−12	972	0.0408	1037
25,000	−30	785	0.0343	1017
30,000	−48	628	0.0288	995
35,000	−66	498	0.0238	973
40,000	−68	392	0.0189	971
45,000	−68	308	0.0148	971
50,000	−68	242	0.0117	971
60,000	−68	151	0.0072	971
70,000	−68	94	0.0045	971
80,000	−68	58	0.0028	971
90,000	−68	36	0.0017	971
100,000	−68	22	0.0011	971
150,000	114	3	9.8×10^{-5}	1174
200,000	159	0.7	2.0×10^{-5}	1220
250,000	−8	0.1	4.8×10^{-6}	1042
500,000	450	10^{-4}	3.1×10^{-9}	—

[†]Reproduced with permission from *Heat, Mass and Momentum Transfer* by W. M. Rohsenow and H. Y. Choi, Prentice-Hall, Inc., Englewood Cliffs, N. J., 1961, p. 523.

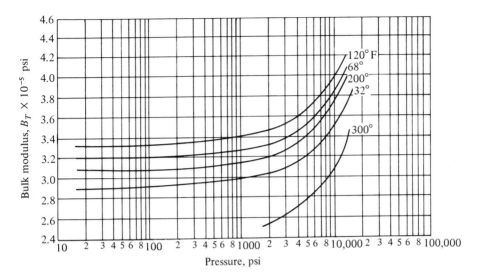

Figure C.1. *Variation of the bulk modulus of water with pressure and temperature. Reproduced with permission from* Fluid Mechanics *by Arthur G. Hansen, John Wiley & Sons, Inc., New York, 1967, p. 487.*

INDEX